Energy Assessments for Industrial Complexes

Authored by

Alexander Spivak,

Ashok Kumar

&

Matthew Franchetti

College of Engineering
The University of Toledo
Toledo, OH 43606
USA

CONTENTS

FOREWORD

The United Nations has defined sustainability as development that meets the needs of the present without compromising the ability of future generations to meet their own needs. In the contemporary business world, sustainability has become much more than an environmental movement, but rather an economic necessity. Between rising costs of raw materials and energy combined with changes in the environmental laws, many businesses are striving to reduce waste and save energy. The benefits of these sustainable actions are reflected in the bottom lines of businesses that embrace sustainability. Green businesses can both save money and appeal to their current and future customers.

The "Energy Assessments for Industrial Complexes" eBook provides a wealth of information to business managers that can help them to improve both environmental and the economic status of their companies. The eBook helps with decision making, such as equipment changes, insulation modifications, and green office management.

The authors of the eBook have a great deal of experience in the environmental field and this contains many practical recommendations and suggestions that can help managers to develop a comprehensive, feasible and financially profitable sustainability program at their facilities.

Christopher Pizza
Coordinator
Lucas County Solid Waste District
Ohio
USA

PREFACE

This eBook discusses methods that businesses may employ to reduce energy costs through environmentally sustainable methods. Environmental sustainability and energy conservation are important topics in today's business climate. There are 23 chapters covering various aspects of energy assessments and each chapter is linked to case histories that are given in the appendix. This eBook discusses how to improve financial and environmental management of most components of a business, from lighting and office supplies to boilers and compressors. Readers will find useful tips on managing their energy needs in a financially sound and environmentally sustainable way. The eBook is designed for mid- and upper-level managers planning to start or currently conducting energy assessments for production sustainability programs for their company.

Those who wish to gain a better understanding of many ways to reduce energy consumption can benefit from this eBook. We are indebted to the former and current Civil Engineering graduate students of the Air Pollution Research Group and Mechanical Industrial and Manufacturing Engineering (MIME) students at The University of Toledo, who were involved in this research over the last 15 years and developed the software that is being used to carry out the energy assessments documented in the Appendix. Special thanks to the Center for Innovative Food Technology (CIFT), TechSolve, Inc., and other participants for of the energy assessments discussed in the appendix of this eBook. We are grateful to the CIFT, TechSolve, Inc., Edison Welding Institute (EWI), and Cleveland Advanced Manufacturing Program (CAMP) for participating in the PPIS grant. We appreciate the research grants awarded by the United States Environmental Protection Agency (U.S. EPA) to The University of Toledo, which made possible the development of software available on our web site and execution of P3 assessments discussed in the Appendices. The contribution made by Srikar Velagapudi during the preparation of Chapter 23 is cheerfully acknowledged. We would also like to thank Kathryn Rose for editorial assistance and her contributions to the structure and format of the eBook.

Even if one user wants to apply energy assessment tools to reduce energy consumption anywhere in the world, this eBook will serve its intended purpose.

These tools will assist environmental professionals in recommending energy strategies for their organization, as well as improving existing tools and developing new approaches. The views expressed in this eBook are those of the authors alone and do not represent the views of the organizations who, over the years, funded the assessments. The authors do not have any financial interest in the software discussed/used in this eBook.

STATEMENT OF CONFLICT OF INTEREST

We acknowledge that our research group has not been paid, sponsored or offered any incentives in any way by the producers mentioned in the eBook.

Alexander Spivak,
Ashok Kumar
&
Matthew Franchetti
College of Engineering
The University of Toledo
Toledo, OH 43606
USA

INTRODUCTION

Alexander Spivak, Ashok Kumar[*] and Matthew Franchetti

University of Toledo, Toledo, OH, USA

Abstract: Introduction to the eBook discusses what is included in the eBook as well as importance of sustainable practices and business need for sustainability. Sustainability analysis tools are listed and discussed.

Keywords: Sustainability, economic sustainability, economy of sustainability, environmental regulations, energy assessment, pollution, pollution prevention, P2, University of Toledo, UT, P2 tools, sustainable practices, MSDS, HVAC, LEAN, HVACDesign, BST, HAT, DGP, SBSAT, D-HAT, CIS, FAT, UT PPIS.

INTRODUCTION

Sustainability, as defined by the EPA, means meeting the needs of the present without compromising the ability of future generations to meet their own needs. The financial equivalent of sustainability would be saving money for the future, as opposed to spending every penny. With changing regulations and perception of environment, sustainability has become an economical, common-sense business practice. Many sustainability practices can both save money and improve image of the company in the eyes of customers and shareholders.

Energy assessments are conducted at business and governmental facilities in order to evaluate current energy practices and derive a comprehensive sustainable energy policy improvement plan. Often, energy assessments are conducted in conjunction with solid waste assessments. Assessments may also include studies of environmental impact, environmental policies, as well as financial impact of the current practices.

The concept of pollution prevention (P2) has been practiced over the last 25 years as a result of the increase in regulatory policies (established by various

---

*Address correspondence to Ashok Kumar:** Department of Civil Engineering, The University of Toledo; 3006 Nitschke Hall, Mail Stop 307, 2801 W. Bancroft St., Toledo, OH 43606, USA; Tel: 419-530-8136, Fax: 419-530-8116; E-mail: ashok.kumar@utoledo.edu

environmental agencies throughout the globe) and their inter-relationship with the concept of sustainability. The United States Pollution Prevention Act of 1990 states that, whenever feasible, pollution should be prevented or reduced at the source. The United States (U.S.) Environmental Protection Agency (EPA) defines P2 as the exercise of practices that reduce or eliminate the creation of pollutants by increased efficiency in the use of raw materials, energy, water, and other resources; and the protection of natural resources by conservation. However, the concept of P2 has merged into sustainability over the last decade in view of the serious environmental challenges, cost competitions, and the ever-increasing consumer demands. Consequently, various environmental agencies and educational institutions have developed several P2 and sustainability tools, including the University of Toledo (UT). UT has developed several user-friendly P2 tools that assist environmental managers determine the opportunity for P2 savings and assess the compliance with regulations. They include the Material Safety Data Sheets tool v1.1/1.2 (MSDS) [1]; the Lean Assessment Screening tool v1.0 (LEAN) [2]; the Hybrid Heating, Ventilation, and Cooling (HVAC) System Design tool v1.0 (HVACDesign) [3]; the Building Sustainability Tool v1.0 (BST) [4]; the Hospital Assessment Tool v1.0 (HAT) [5]; the Database for Green Products v1.0 (DGP) [6]; the Small Business Self-Assessment Tool v1.0 (SBSAT) [7]; the Department Specific Hospital Assessment Tool v1.0 (D-HAT) [8]; Chemical Identification Software v1.0 (CIS) [9]; and the Food Assessment Tool v1.0 (FAT) [10];. These tools are available online to environmental managers, in assessing their opportunities for P2 savings, and are free to download from the University of Toledo web site (UT PPIS) Website. Available from http://www.eng.utoledo.edu/aprg/ppis/ppistools.htm (2012) [11]. They provide a comprehensive outlook into the various aspects of P2 activities. Kumar *et al.* (2012) [12] discusses the use of the Economic Input Output Life Cycle Assessment (EIO-LCA) tool to determine the change in environmental impact after implementation of the recommendations in association with the use of the five P2 tools for automobile parts manufacturing facility located in Findlay, Ohio.

This eBook focuses on different sustainability practices, discussion of their financial implications, and related energy assessments. Included material discusses how to apply sustainability principles to generate immediate and long-term savings as well as public goodwill for the business. We discuss sustainable projects in multiple areas for general business including illumination, windows, doors, insulation, appliances, computers and office equipment, HVAC systems, vending machines, roofs, water heaters, documents, and solid waste management. We include chapter's specific to manufacturing such as boilers, insulation, motors, garage doors, plant floor machinery, compressors, belt conveyors, and heat recovery. Chapters also include special topics such as showers, pools and kitchen hoods. The eBook concludes with a chapter on application of calculators for the energy assessment and examples of related energy assessments.

REFERENCES

[1] Kumar A, D'Souza F, Vashisth S, Software for Material Safety Data Sheets. Environ. Prog. 1996; 15- 2: S17-S23.

[2] Kumar A, Thomas S. A Software Tool for Screening Analysis of Lean Practices. Environ. Prog. 2002; 21- 3: O12-O16.

[3] Pendse R, Kumar A, Vijayan A. Development of a Spreadsheet to Determine Natural Ventilation Cooling Hours for a Commercial Hybrid HVAC System. Environ. Prog. 2005; 24-1: 16-23.

[4] Vijayan A, Kumar A. Development of a Tool for Analyzing the Sustainability of Residential Buildings in Ohio. Environ. Pro. 2005; 24-3: 238-247.

[5] Raman N, Vijayan A, Kumar A. Development of a Pollution Prevention Tool For Assessment of Hospital Waste. Environ. Pro. 2006; 25-2: 93-98.

[6] Nimse P, Vijayan A, Kumar A, Varadarajan C. A Review of Green Product Databases, Environ. Pro. 2007; 26-2: 131-137.

[7] Kadiyala A, Kumar A. Development of an Environmental Compliance Tool for Small Businesses. Environ. Pro. 2007; 26-4: 316-326.

[8] Kadiyala A, Somuri D, Kumar A. Development of a Tool for the Assessment of Department Specific Hospital Waste. Environ. Pro. 2008; 27-4: 432-438.

[9] Kadiyala A, Velagapudi S, Kumar A. Development of Chemical Identification Software for Multi-National Industries. Environ. Prog. Sustainable Energy 2009; 28-1:13-19.

[10] Kadiyala A, Nerella VVK, Kumar A. Development of an Assessment Tool for Pollution Prevention and Energy Efficiency in Food Industry. Environ. Prog. Sustainable Energy. 2009; 28-3: 310- 315.

[11] University of Toledo web site (UT PPIS) Website. Available from http://www.eng.utoledo.edu/aprg/ppis/ppistools.htm (2012).

[12] Kumar A, Velagapudi S, Kadiyala A. Energy Assessments Using the Pollution Prevention Tools. In: Cavalcanti EFS and Barbosa MR, Eds. Energy Efficiency: Methods, Limitations and Challenges, NY: Nova Science Publishers, Inc. 2012; pp. 169-182.

Energy Assessments for Industrial Complexes

Send Orders of Reprints at reprints@benthamscience.net

CHAPTER 1

Optimization of Lighting

Alexander Spivak, Ashok Kumar[*] and Matthew Franchetti

University of Toledo, Toledo, OH, USA

Abstract: This chapter will help optimize the amount of lighting required in a specific room. Discussed methods and calculations will help managers to estimate if excessive lighting is currently being used and how much money may be saved by removing excessive lighting. In addition, this chapter will address a case of insufficient lighting.

Keywords: Lighting, lighting solutions, minimum lighting requirement, lighting replacement decisions, energy savings, energy consumption, efficacy, illumination, lumens, lux, lighting in the facility, required lighting, lighting fixtures, light bulb, bulb type, CRI (Color Rendering Index), CCT, correlated color temperature, automatic light switch.

INTRODUCTION

Lighting updates is among the most common recommendations of the energy assessments. With relatively low cost and ease of implementation, lighting is one of the "low hanging fruits" that lead to substantial energy savings. Here we will introduce various energy savings possibilities and constraints associated with lighting choice.

MINIMUM LIGHTING REQUIREMENTS

Lighting requirements vary depending on facility use. As visual demands of tasks increase, so does the number of required lumens in the area where tasks are performed increase. Facilities must have sufficient amount of lumens per square foot, but not necessarily excessive amount of light. Lighting requirements may be met by strategically placing light fixtures and by selecting light fixtures with

*Address correspondence to Ashok Kumar:** Department of Civil Engineering, The University of Toledo; 3006 Nitschke Hall, Mail Stop 307, 2801 W. Bancroft St., Toledo, OH 43606, USA; Tel: 419-530-8136, Fax: 419-530-8116; E-mail: ashok.kumar@utoledo.edu

lowest feasible total cost of ownership. Table **1** describes required lighting for various facilities.

Table 1: Required Illumination Chart [1]

Activity	Illumination (Lux, Lumens/m^2)
Public areas with dark surroundings	20 - 50
Simple orientation for short visits	50 - 100
Working areas with occasional visual tasks	100 - 150
Archives, homes, theaters, warehouses, *etc.*	150
Classrooms, limited office work	250
Computer work, groceries, laboratories, regular office work, show rooms, study areas	500
Mechanical workshops, office landscapes, supermarkets	750
Detailed mechanical workshops, normal drawing workshops, operation theaters	1,000
Very detailed mechanical workshops, detailed drawing workshops	1,500 - 2,000
Prolonged visual tasks at low contrast and very small size (*e.g.,* some surgeries)	2,000 - 5,000
Very prolonged exacting visual tasks (*e.g.,* surgeries, jewelry and watch making)	5,000 - 10,000
Very special visual tasks of extremely low contrast and small size	10,000 - 20,000

The list of activities in the chart presented is not inclusive. However, a facility may extrapolate its illumination needs from this chart. For example, a facility may have offices, mechanical workshops, warehouse, *etc.* Each area, if possible, can receive its appropriate illumination: 500, 750, 150 lux respectively. However, within areas, there may be a section where additional lighting is required. For example, an area in a mechanical workshop where a quality controller is required to regularly check very small parts may require additional (perhaps local) illumination.

ESTIMATION OF REQUIRED LIGHTING

Calculation of the number of lumens at each location of a room is somewhat complex and falls outside of our scope. Determining these factors is not simple or practical in existing facilities, but can make a significant difference in newly built facilities.

A more effective and easy (but somewhat crude) option is to use a light meter that is able to operate within desired range of lumens or foot-candles (note: 1 lux = 1 lumen/m^2 = 0.0001 phot = 0.0929 fcd (foot candle) = 0.0929 lumen/ft^2). A simple light meter can cost as little at $80 and is very easy to operate.

If average light meter measurements determine that the amount of light is not consistent with general recommendations in Table **1**, the selection of lighting may lead to following questions:

- Should some light fixtures (luminaries) be disconnected and otherwise relocated?

- Should some light fixtures (luminaries) be disconnected and light bulbs and/or luminaries replaced with more efficient ones?

Total cost savings from changes to existing fixtures can be calculated by formula:

$$S = H \times W \times (R_r - R_e) \times C \tag{1}$$

Where:

S is cost savings due to changes to existing fixtures

H is a hours per day existing fixture is in operation

W is workdays per year

R_r is total kW-hr rating of all replacement fixtures

R_e is total kW-hr rating of all existing fixtures

C is cost of electricity per kW

where the cost of the electrical updates may be obtained from the estimates of electrical contractors or electricians employed by the company.

The total payback period may then be estimated by calculating how many years will it take to offset the cost of electrical updates. The payback period may be adjusted if facility is implementing programs that reduce number of hours of lights being on by installation of motion sensors and other equipment discussed later in the chapter.

REDUCTION OF COSTS AND SUSTAINABILITY IMPROVEMENT THROUGH DISCONNECTION/RELOCATION OF LUMINARIES

A decision to reduce number of light fixtures is best done by replacing light bulbs with non-working ones, keeping off fixtures that can be controlled by a separate switch or breaker, or by disconnecting or bypassing the fixture and capping wires for easy re-connection if needed. After fixtures are disconnected or relocated, a light meter and employee opinion may help to determine if amount of lighting is sufficient to perform job-related tasks. Local light fixtures may be used to supplement specific area(s), such as specific assembly stations.

To improve mobility of the lights in the future, use fixtures with a 6-foot pigtail so they can be moved in the future as floor layout changes or additional equipment is installed.

REDUCTION OF COSTS AND SUSTAINABILITY IMPROVEMENT THROUGH IMPROVING LIGHTING EFFICACY

In conjunction with disconnecting and relocating lights, it may be practical to replace light fixtures that generate greater amount of light while consuming less electricity. A common recommendation aimed at reducing the overall electrical bill due to lighting is to change from T-12 lights to T-8 or T-5 light fixtures.

Numeric value of the fluorescent light refers to diameter of the lamp's tube. For example, T-8's tube diameter is $\frac{8}{8} in$ (or 1 inch), and T-12's tube diameter is $\frac{12}{8} in$ (or 1.5 in). The reduction of the diameter allows for greater energy efficiency. In fact, amount of light generated by T-8 and T-12 lights is about the same, while T-8 uses approximately 20% less energy.

Lumens output and efficiency of various types of popular lighting is presented in Table **2** [1-7].

Table 2: Lumens Output and Efficacy Chart.

Bulb Type Examples[†]	Initial Lumens Generated	Average Lumens Generated*	Watts Used**	Efficacy (Lumens/Watts)	T-12 Intensity Multiplier
Incandescent (neutral)		1210	75	16.1 (from 7.4 to 21.0)	0.22
T-12 Normal Output (cool)	3200	2880	40	72.0	1.00
T-12 High Output (cool)	4050	3281	60	54.7	0.76
T-12 Very High Output (cool)	6600	4620	115	40.2	0.56
T-8 Nomal Output (cool)	2800	2520	32	78.8	1.09
T-5 Normal Output (cool @ 35C)	2900	2755	28	98.4	1.37
T-5 High Output (cool @ 35C)	5000	4750	54	88.0	1.22
Spiral Compact		1250	20	62.5	0.87
Power Compact	4800	4128	75	55.0	0.76
Metal Halide		40,000	400	100.0 (from 80 to 120)	1.39
High Pressure Sodium		50,000	400	125.0 (from 50 to 140)	1.74
LED (prototype)				32	0.44

Bulb Type Examples	T-12 Replacement Quantity	Life Expectancy	Approximate Cost	CRI Index
Incandescent (neutral)	5 : 2	500 - 1000 hours	low	99+
T-12 Normal Output (cool)	1 : 1	7,000 - 20,000 hours***	medium	62 - 89
T-12 High Output (cool)	1 : 1	7,000 - 20,000 hours***	medium	62 - 89
T-12 Very High Output (cool)	3 : 5	7,000 - 20,000 hours***	medium	62 - 89
T-8 Nomal Output (cool)	1 : 1	7,000 - 20,000 hours***	medium	62 - 89
T-5 Normal Output (cool @ 35C)	1 : 1	7,000 - 20,000 hours	medium	62 - 89
T-5 High Output (cool @ 35C)	3 : 5	7,000 - 20,000 hours	medium	62 - 89
Spiral Compact	7 : 3	6,000 - 10,000 hours	medium	60 - 90
Power Compact	2 : 3	6,000 - 10,000 hours	medium	60 - 90
Metal Halide	1 : 14	5,000 - 20,000 hours	high	65 - 92
High Pressure Sodium	1 : 17	16,000 - 24,000 hours	high	22 - 70
LED (prototype)		claim: 50,000 hours	high	

†List of light fixtures is not inclusive. Mercury vapor, Low pressure sodium, *etc.* light bulbs were not included.
*Over time, T-5s loose lumens slower than T-8s, which loose lumens slower than T-12s.
**A single wattage was used in the table. All bulbs are available with various levels of wattage. For example, incandescent bulbs range from 5 to 300 watts (to be phased out by 2014 in US due to low efficacy), compact lights range from 5 to 125 watts. The higher the wattage the greater number of lumens is generated.
*** Prolonged life units are rated up to 30,000 hours.

Table **2**'s efficacy comparison may help to determine which lighting fixtures are most economical. The system's efficacy is related to but differs from light's efficacy. When selecting lighting systems, rated efficacies of specific units are available from manufacturers.

One has to be careful not to make decisions on economic measures alone (efficacy being the greatest economic factor - cost of energy can be as high as 95% of the total cost of lighting), but consider other factors such as temperature, age, lamps, life expectancy, disposal costs, and quality. For example, selection of insufficient illumination may result in an injury.

The efficacy is calculated by dividing average lumens produced into watts consumed. The average number of lumens produced by the light fixture is factor of a number of conditions, which include temperature of operation and wiring. For example, light output at 30°C is about the same for T-5 and T-8, however T-8 will perform relatively better at lower ambient temperatures and T-5 will perform relatively better at higher ambient temperatures.

Age of the light bulb is another factor that affects light output is. All fluorescents have a "burn-in" period, during which they are much brighter. The side-by-side comparisons of lamps should is made after they have passed their "burn-in" period.

Lights dim overtime. For example, T-5 retains 95% of lumens, T-8–88%, and T-12–73%[6]. The difference is mainly due to the type of ballast: magnetic *vs.* electronic.

T-12 replacement quantity table is a general guideline. For example, five T-12 bulbs may be replaced with three high output T-5 bulbs. This does not mean that replacement may be completed directly. T-5 lamps cannot replace T-12 or T-8 lamps because of differences in physical and electrical characteristics. T-5 lamps are slightly shorter than T-12 lamps and therefore cannot be used as replacements for the larger lamps without special adjustments (except for luminaries which are designed for both lamps). In addition, T-5 lamps have different electronic ballasts. However, tube converters are now available for such replacements. Ballasts available for lights are matched to specific application. Typical ballasts used are magnetic, electronic and cathode-disconnect.

The quantity of light and color the light generated are controlled by light's ballasts. Magnetic ballasts used with T-12 lights have ballast factor lower than 100%. It means that lamp is able to produce less than 100% of its design lumens. However, electronic ballasts on T-8 lights have ballast factors of greater than 100%, making them even more efficient compared to T-12 lights.

Life expectancy of the bulb depends on multiple factors, such as wiring/voltage, proper ballast, temperature in the facility and number of starts. Since life of lamps varies significantly, it is hard to predict the life expectancy of the unit lamp. However, since price of the unit bulb and replacement labor accounts for less than 1% of the total electrical cost (in typical case, a new lamp will cost $10 - $30 installed), a conservative value for the life expectancy should be used in calculations. The new luminary may cost between $100 and $500 installed.

The disposal cost of the light bulb is also very small, unless manufacturer has excessive quantity of mercury generated from other processes. Example mercury (Hg) contents are: T-5 HO contains 3 mg HG, T-8 NO contains 15 mg Hg, T-12 NO contains between 11 and 30 mg Hg, and Metal Halide bulb contains between 40 and 65 mg Hg [6]. Green tip lights are generally a good idea to consider. However, it is important to remember that they still contain mercury and are considered hazardous waste unless proven otherwise through rather expensive testing or confirmed to be non-hazardous by the manufacturer.

Replacing T-12 with T-8 bulbs usually results in a five to seven year payback period in a commercial setting [7]. If all lamps cannot be upgraded simultaneously, upgrading as old T-12 lamps burn out is another option. However, such option does not allow for effective strategic relocation of lights.

CRI (Color Rendering Index) is an important measure of the quality of light. At CRI = 100, colors appear as they would under daylight spectrum at the same CCT vales. Values of CRI less than 50 are generally not acceptable, because colors no longer appear natural. CRI is measured at specific CCT (Correlated Color

Temperature), which describes dominant color tone (very cool–5,000K, cool–4,100K, neutral–3,500K, and warm–3,000K). An increase in CRI improves the actual and perceived quality of the lighting in the work area. Generally as CRI index increases, lumens output decreases.

SIMPLE LIGHTING OPTIMIZATION METHOD

Lighting optimization is a technique of placing lights in such a way so that minimum required amount of light is available in every part of the work area. The amount of lighting required in each specific job site is shown in Table **2**. A simple way to estimate total lighting may be done by dividing 120% of the total amount of lighting required by the amount of lumens generated by each light fixture. The quotient, rounded up, is number of light fixtures needed in the work area. The additional 20% of the lighting requirement is an acceptable way to compensate for reduced performance of lighting fixtures over time.

$$N = \frac{1.2 \times T}{F} \tag{2}$$

Where:

   N is required number of light fixtures

   T is amount of lighting required (in lumens)

   F is lumens generated by each lighting fixture

Fixtures should then be evenly distributed across the ceiling area. Standard light fixture distribution patterns are available; however common sense approach may work better for specific applications. As mentioned above, light fixtures with long pigtails may be more easily moved as needed.

If some areas of the room require more lighting than others, additional lighting may be used at or above the workstation for greater illumination. The decision is easy to make, since majority of the cost is usually due to electrical consumption, not the cost of light fixtures. However, if light fixtures are expensive, payback calculations (as shown earlier in the chapter) may be used.

OTHER SUSTAINABLE LIGHT/ENERGY SAVING METHODS

The most obvious way to save energy is by switching off lighting when not in use. There are two general approaches: technologically and manually. Experience suggests that technological approach is much more effective in most cases. This approach entails motion sensors and automatic timers in offices, meeting rooms, bathrooms and (something that is typically not done) some manufacturing areas, such as paint rooms, areas below main floor level, *etc.* The payback is easy to calculate: it is cost of installation of equipment divided by estimated cost of electricity saved. Typical payback periods are short.

The alternative method is manually turning of lights. It is not as effective due to human nature, though standard operating procedures can be implemented. At times, tier I suppliers at large manufacturing facilities take charge of shutting off lights at some manufacturing areas when not in use. They report it as part of cost aversions they must generate every year.

In addition to lighting, other pieces of equipment may be turned off either automatically or as part of SOP. For example, if production line requires fans or blowers during production, wiring fans to stop ten minutes after production conveyor stops may save significant amount of money on electricity. However, one should be careful with this approach during production day due to ramp-up curve electrical consumption issues. These techniques will be discussed further in the eBook.

During the holidays many retail businesses attract customers by addition of decorative lighting, using old incandescent lights. However, a switch to LED lights may pay back in as little as one year. To calculate payback period, the company needs to divide the cost of purchase of new lighting by the cost of electricity consumed. Cost of electricity is the difference kW-hr of the two lighting systems multiplied by the cost of electricity per kW-hr multiplied by the number of hours of operation.

CONCLUSIONS

These are simple techniques that may be used to sustainably reduce energy costs due to lighting. The simple payback calculations recommended in this chapter must include all upfront costs, such as costs of luminaries and installation, and energy cost savings due to changes in the lighting. Managers must always approach changes with caution, in order to avoid potential problems that may result in a decrease of production or increase in injuries.

Note: Related assessments found in appendix are: A, B, C, D, H, I, J, M, O, R, V, W.

REFERENCES

[1]   The Engineering Toolbox website. Available from: www.engineeringtoolbox.com/light-level-rooms-d_708.html. [Accessed March 2012].

[2]   The Planted Tank website. Available from: www.plantedtank.net/articles/Light-Bulb-Comparison/29/. [Accessed March 2012].

[3]   Good Mart website. Available from: http://goodmart.com/light_bulbs_ballasts.htm. [Accessed March 2012].

[4]   GrowCo Indoor Garden Supply website. Available from: www.4hydroponics.com. [Accessed March 2012].

[5]   Specialty-lights website. Available from: www.specialty-lights.com/plant-grow-faq3.html. [Accessed March 2012].

[6]   Lighting Research Center website. Available from: www.lrc.rpi.edu. [Accessed March 2012].

[7]   Phillips I. Fluorescent Lamps and Ballasts [Steve's Shop O'Sawdust website]. Available from: http://wood.bigelowsite.com/articles/fluorescent_lamps_and_ballasts.htm. [Accessed March 2012].

Send Orders of Reprints at reprints@benthamscience.net
Energy Assessments for Industrial Complexes, 2013, 13-18

CHAPTER 2

Windows

Alexander Spivak, Ashok Kumar[*] and Matthew Franchetti

University of Toledo, Toledo, OH, USA

Abstract: Overall heat loss from the building is a sum of the losses of heat through windows, doors, walls, roof, floors, basements, ventilation systems and infiltration. This chapter specifically discusses heat loss through windows and sustainability techniques that help improve bottom line.

Keywords: Window, window insulation, window replacement, U-value, R-value, payback period, heat transfer, heat loss, cost of heating, cost of cooling, old window, draft, draft cost, draft cost calculation, heat loss, heat loss across window, type of window, window frame, argon gas effect, when to replace window.

INTRODUCTION

There are two ways of reducing the heat loss through the window: reduction or elimination of drafts and window replacement. The payback period of the former is immediate, while the latter's payback period depends on the replacement of window's qualities such as R-values, cost of heat in the facility, size and location of the window, *etc.* as well as condition of the window to be replaced. The chapter will discuss estimation of replacement payback periods for window replacements.

COST OF DRAFTS

The first and most obvious step to reduce heat loss though windows is to test windows for draft by placing a moist hand near the edges of the window, using flickering candle, forcing air through using colored gas or dust, or using any other technique Any drafts felt may be sealed off with a window and door-rated caulk

---

*Address correspondence to Ashok Kumar: Department of Civil Engineering, The University of Toledo; 3006 Nitschke Hall, Mail Stop 307, 2801 W. Bancroft St., Toledo, OH 43606, USA; Tel: 419-530-8136, Fax: 419-530-8116; E-mail: ashok.kumar@utoledo.edu

(with silicone) [1, 2]. In the areas where caulk may not be applied, such as parts of the windows that are designed to open, insulation inserts may be used. The payback can be calculated by multiplying specific heat of air by the amount of air escaped per unit time by the difference in temperature, as shown below [3]:

$$E = C \times V \times (T_2 - T_1)$$ (1)

Where:

E is energy loss through opening (in Watts)

C is specific heat of air (in J/cm^3-K)

V is amount of air escaped (in cm^3/day)

T_2 is interior temperature (in degrees C or K)

T_1 is exterior temperature (in degrees C or K)

WINDOW'S REPLACEMENT COST

A window's surface can be another major source of the heat loss. The heat loss may be estimated by knowing the area of the window, temperature difference (average daily value for the year x 365) and U-value (thermal transmittance coefficient) of the window. The U-values for new windows are available from the manufacturer. Typically, U-values of new windows are at or below 0.35. The U-value of a window is the rate of loss of heat in watts per square meter of that window per degree centigrade temperature difference across the window. The lower the U-value, the greater is the window's insulating value. U-value is 1/R-value. Therefore,

$$E = A \times Q \times (T_2 - T_1)$$ (2)

Where:

E is energy loss through window glass (in Watts)

A is area of the window (in m^2)

Q is window's coefficient (in W/m^2-K)

T$_2$ is interior temperature (in degrees C or K)

T$_1$ is exterior temperature (in degrees C or K)

U-VALUE ESTIMATION

In order to estimate U-value of an older window, a thermal gun can determine the temperatures on both sides of the window. Then, the temperature difference can be plugged into the equation above. Alternatively, an estimate of the U-value may be made from Tables **1-3** presented further in the text.

Table 1: Basic U-Value Ratings by Window Type [4, 5]. More Detailed Information is Available Below:

Window type	Estimated U-Value
Old metal casement window	1.30
Good quality single pane window	1.00
Single pane (wood, non metal frame)	0.85
Good quality single pane window with storm	0.60
Double pane with low-E glass	0.40
Double glazed, argon gas (nonmetal frame)	0.28
Triple pane with low-E glass	0.25

Table 2: Window U-Value Estimates [6]

Descruotion[1,2,3,4]			Frame Type[5,6]		
			Aluminum	Aluminum Thermal Break[7]	Wood/Vinyl
Windows	Single		1.20	1.20	1.20
	Double, < 1/2"	Clear	0.92	0.75	0.63
		Clear + Argon	0.87	0.71	0.60
		Low-e	0.85	0.69	0.58
		Low-e + Argon	0.79	0.62	0.53
	Double, ≥ 1/2"	Clear	0.86	0.69	0.58
		Clear + Argon	0.83	0.67	0.55
		Low-e	0.78	0.61	0.51

Table 2: contd....

			0.75	0.58	0.48
	Triple	**Low-e + Argon**	0.75	0.58	0.48
		Clear	0.70	0.53	0.43
		Clear + Argon	0.69	0.52	0.41
		Low-e	0.67	0.49	0.40
		Low-e + Argon	0.63	0.47	0.37
Garden Windows	**Single**		2.60	n/a	2.31
	Double	**Clear**	1.81	n/a	1.61
		Clear + Argon	1.76	n/a	1.56
		Low-e	1.73	n/a	1.54
		Low-e + Argon	1.64	n/a	1.47

[1]: <1/2" = a minimum dead air space of less than 0.5 inches between the panes of glass, >1/2" = a minimum dead air space of 0.5 inches or greater between the panes of glass, Where no gap width is listed, the minimum gap width is 1/4 ‖ .
[2]: Any low-e (emissivity) coating (0.1, 0.2 or 0.4).
[3]: U-factors listed for argon shall consist of sealed, gas-filled insulated units for argon, CO_2, SF_6, argon/SF_6 mixtures and Krypton.
[4]: "Glass block" assemblies may use a U-factor of 0.51.
[5]: Insulated fiberglass framed products shall use wood/vinyl U-factors.
[6]: Aluminum clad wood windows shall use the U-factors listed for wood/vinyl windows.
[7]: Aluminum Thermal Break = An aluminum thermal break framed window shall incorporate the following minimum design characteristics:
    a)   The thermal conductivity of the thermal break material shall be not more than 3.6 Btu-in/h/ft2/°F;
    b)   The thermal break material must produce a gap in the frame material of not less than 0.210 inches; and,
    c)   All metal framing members of the products exposed to interior and exterior air shall incorporate a thermal break meeting the criteria in a) and b) above.

Table 3: Window U-Value Estimates [6]

Vertical Glazing Description				Frame Type		
Panes	**Low-e[1]**	**Spacer**	**Fill**	**Any Frame**	**Aluminum Thermal Break[2]**	**Wood/Vinyl/Fiberglass**
Double[3]	A	Any	Argon	0.48	0.41	0.32
	B	Any	Argon	0.46	0.39	0.30
	C	Any	Argon	0.44	0.37	0.28
	C	High performance	Argon	0.42	0.35	Deemed to comply[5]
Triple[4]	A	Any	Air	0.50	0.44	0.26
	B	Any	Air	0.45	0.39	0.22
	C	Any	Air	0.41	0.34	0.20
	Any double low-e	Any	Air	0.35	0.32	0.18

[1]: Low-eA (emissivity) shall be 0.24 to 0.16, Low-eB (emissivity) shall be 0.15 to 0.08 and Low-eC (emissivity) shall be 0.07 or less.
[2]: Aluminum Thermal Break = An aluminum thermal break framed window shall incorporate the following minimum design characteristics:

a) The thermal conductivity of the thermal break material shall be not more than 3.6 Btu-in/h/ft2/°F;

b) The thermal break material must produce a gap in the frame material of not less than 0.210 inches; and

c) All metal framing members of the products exposed to interior and exterior air shall incorporate a thermal break meeting the criteria in a and b above.

[3]: A minimum air space of 0.375 inches between panes of glass is required for double glazing.

[4]: A minimum air space of 0.25 inches between panes of glass is required for triple glazing.

[5]: Deemed to comply glazing shall not be used for performance compliance.

ESTIMATION OF PAYBACK PERIOD FOR WINDOW REPLACEMENT

The payback period may now be estimated by calculating saved wattage and converting it into annual energy savings by multiplying the cost of 0.001 kW-hr by 8760 hours per year. Savings information has to be taken over the period of a year, since windows that perform well to preserve heat may not be as efficient saving on air conditioning and *vice versa*, as has been shown [5]. "Perfect window" [6] study demonstrated desired characteristics of "perfect summer window" and "perfect winter window". Window glass characteristics discussion is beyond the scope of the eBook. However, readers in hot and cold climates are encouraged to learn about and consider windows best suited for their area's climatic conditions. Then savings may be divided into initial investment of purchasing and installing windows to determine how many years will it take to pay back the initial investment. Depending on window size, cost of utilities, incentives, location, and condition of current windows a window replacement could have anywhere from a 10 to 50-year ROI [3]. For example, a detailed study of effects of window type upon total building energy cost in British Columbia, Canada showed that non-metal frame low-e argon filled triple-e windows may save over 8% of energy costs [4].

CONCLUSIONS

The most cost effective method to reduce heat loss through windows is to eliminate the drafts around windows. The next step is to consider replacing windows with high efficiency windows. Typically, older windows and single-pane windows are best candidates for replacement; however, even newer windows may be considered for replacement.

Note: Related assessments found in appendix are: B.

REFERENCES

[1]   The Energy Toolbox website. Available from: http://www.engineeringtoolbox.com/heat-loss-buildings-d_113.html. [Accessed March 2012].

[2]   Becker, P. Technical paper #10: U-values and Traditional Buildings: *In Situ* Measurements and their Comparison to Calculated Values. [Historic Scotland Alba Aosmhor website]. Available from: http://www.historic-scotland.gov.uk/hstp102011-u-values-and-traditional-buildings.pdf. [Accessed March 2012].

[3]   US Green Building Council's Green Home Guide website. Available from: http://greenhomeguide.com/askapro/question/is-there-a-way-to-measure-u-values-on-older-windows-to-determine-if-replacement-makes-sense. [Accessed March 2012].

[4]   Pape-Salmon A, Knowles W. Transforming the Window and Glazing Markets in BC through Energy Efficiency Standards and Regulations. ASHRAE Transactions 2011; 117-1: 306-313.

[5]   Masca S, Yasar Y. The Effects of Window Alternatives on Energy Efficiency and Building Economy in High-Rise Residential Buildings in Cold Climates. GUJ Sci 2011; 24-4: 927-944.

[6]   Ye H, Meng X, Xu B. Theoretical discussions of perfect window, ideal near infrared solar spectrum regulating window and current thermochromic window. Energy and Building 2012; 49: 164-172.

[7]   The U-factor of Thermal Replacement Windows. [Servicemagic website]. Available from: http://www.servicemagic.com/article.show.The-U-factor-of-Thermal-Replacement-Windows.8839.html. [Accessed March 2012].

[8]   A Short Course in Windows. [Village of Glendale, OH website]. Available from: http://www.glendaleohio.us/PDF/Short_Course_in_Windows.pdf. [Accessed March 2012].

[9]   Washington State Energy Code 2009 Edition Chapter 51-11 WAC. [Kitsop County website].   Available   from: http://www.kitsapgov.com/dcd/documents/WSEC%2009%20Res%20Amendments.pdf. [Accessed March 2012].

Send Orders of Reprints at reprints@benthamscience.net
Energy Assessments for Industrial Complexes, 2013, 19-24

CHAPTER 3

Doors

Alexander Spivak, Ashok Kumar* and Matthew Franchetti

University of Toledo, Toledo, OH, USA

Abstract: This chapter will discuss heat loss through doors, which is reasonably similar to heat loss through windows discussed in Chapter 3, and sustainability techniques that help improve bottom line.

Keywords: Doors, door insulation, U-value R-value, payback period, heat transfer, heat loss, cost of heating, cost of cooling, old door, draft, draft cost, draft cost calculation, heat loss, heat loss across door, type of door, door design, door shape effect, when to replace door.

INTRODUCTION

Just like windows, the main consideration for doors is how much money can be saved in energy costs if the door is to be replaced. The U-values in this chapter may be applied to equations in Chapter 3 to estimate a payback period for a new door.

ESTIMATION OF DOOR'S U-VALUE

Door U-values (U-factors) are calculated in the same way as window U-values. Below are the tables (Tables **1-6**) for default u-factors of currently installed doors.

Table 1: Commercial Vertical Glazing, Overhead Glazing and Opaque Doors [1]

Vertical Glazing	U-Factor		
	Any Frame	Aluminum with Thermal Break	Wood/Vinyl/Fiberglass Frame
Single	1.45	1.45	1.45

*Address correspondence to Ashok Kumar: Department of Civil Engineering, The University of Toledo; 3006 Nitschke Hall, Mail Stop 307, 2801 W. Bancroft St., Toledo, OH 43606, USA; Tel: 419-530-8136, Fax: 419-530-8116; E-mail: ashok.kumar@utoledo.edu

Table 1: contd….

Double	0.90	0.85	0.75
1/2" air, fixed/operable	0.75/0.90	0.70/0.84	0.60/0.72
1/2" air, low-e (0.40), fixed/operable	0.70/0.84	0.60/0.72	0.50/0.60
1/2" air, low-e (0.10), fixed/operable	0.65/0.78	0.44/0.66	0.45/0.54
1/2" argon, low-e (0.10), fixed/operable	0.60/0.72	0.50/0.60	0.40/0.48
Triple	0.75	0.55	0.50
1/2" air, fixed/operable	0.55/0.66	0.50/0.60	0.45/0.54
1/2" air, low-e (0.20), fixed/operable	0.50/0.60	0.45/0.54	0.40/0.48
1/2" air, 2 low-e (0.10), fixed/operable	0.45/0.54	0.35/0.42	0.30/0.36
1/2" argon, 2 low-e (0.10), fixed/operable	0.40/0/48	0.30/0.36	0.25/0.30

Overhead Glazing: Sloped Glazing (including frame)	**U-Factor**		
	Any Frame	**Aluminum with Thermal Break**	**Wood/Vinyl/Fiberglass Frame**
Single	1.74	1.74	1.74
Double	1.08	1.02	0.90
1/2" air, fixed	0.90	0.84	0.72
1/2" air, low-e (0.40), fixed	0.84	0.72	0.60
1/2" air, low-e (0.10), fixed	0.78	0.66	0.54
1/2" argon, low-e (0.10), fixed	0.72	0.60	0.48
Triple	0.90	0.66	0.60
1/2" air, fixed	0.66	0.60	0.54
1/2" air, low-e (0.20), fixed	0.60	0.54	0.48
1/2" air, 2 low-e (0.10), fixed	0.54	0.42	0.36
1/2" argon, 2 low-e (0.10), fixed	0.48	0.36	0.30

Opaque Doors	**U-Factor**
Uninsulated metal	1.20
Insulated metal (incl. fire door and smoke vent)	0.60
Wood	0.50

NOTES:
Where a gap width is listed (*i.e.*, 1/2 inch), that is the minimum allowed.
Where a low-emissivity emittance is listed (*i.e.*, 0.40, 0.20, 0.10), that is the maximum allowed.
Where a gas other than air is listed (*i.e.,* Argon), the gas fill shall be a minimum of 90%.
Where an operator type is listed (*i.e.*: Fixed), the default is only allowed for that operator type.
Where a frame type is listed (*i.e.,* Wood/Vinyl), the default is only allowed for that frame type. Wood/Vinyl frame
Includes reinforced vinyl and aluminum-clad wood

Table 2: Residential Doors [1]

Door Type: Swinging Doors (Rough Opening 38" x 82")	No Glazing	Single Glazing	Double Glazing with 1/4" Airspace	Double Glazing with 1/2" Airspace	Double Glazing with e=0.10, 1/2" Argon
Slab doors					
Wood slab in wood frame[a]	0.46				
6% glazing (22" x 8" lite)	--	0.48	0.47	0.46	0.44
25% glazing (22" x 36" lite)	--	0.58	0.48	0.46	0.42
45% glazing (22" x 64" lite)	--	0.69	0.49	0.46	0.39
Insulated steel slab with wood edge in wood frame[a]	0.16				
6% glazing (22" x 8" lite)	--	0.21	0.20	0.19	0.18
25% glazing (22" x 36" lite)	--	0.39	0.28	0.26	0.23
45% glazing (22" x 64" lite)	--	0.58	0.38	0.35	0.26
Foam insulated steel slab with metal edge in steel frame[b]	0.37				
6% glazing (22" x 8" lite)	--	0.44	0.42	0.41	0.39
25% glazing (22" x 36" lite)	--	0.55	0.50	0.48	0.44
45% glazing (22" x 64" lite)	--	0.71	0.59	0.56	0.48
Cardboard honeycomb slab with metal edge in steel frame[c]	0.61				
Site-assembled style and rail doors					
Aluminum in aluminum frame	--	1.32	0.99	0.93	0.79
Aluminum in aluminum frame with thermal break	--	1.13	0.80	0.74	0.63

[a]: Thermally broken sill (add 0.03 for non-thermally broken sill).
[b]: Non-thermally broken sill.
[c]: Nominal U-factors are through the center of the insulated panel before consideration of thermal bridges around the edges of the door section and due to the frame.

Table 3: Emergency Exit Door [1]

Double-Skin Steel Emergency Exit Doors			
Core Insulation		3' x 6' 8"	6' x 6' 8"
1-3/8" thickness	Honeycomb kraft paper	0.57	0.52
	Mineral wool, steel ribs	0.44	0.36
	Polyurethane foam	0.34	0.28
1-3/4" thickness	Honeycomb kraft paper	0.57	0.54

Table 3: contd....

	Mineral wool, steel ribs	0.41	0.33
	Polyurethane foam	0.31	0.26

Table 4: Revolving Doors[1]

Revolving Doors		
	Size (W x H)	**U-Factor**
3-wing	8' x 7'	0.79
	10' x 8'	0.8
4-wing	7' x 6.5'	0.63
	7' x 7.5'	0.64
Open	82" x 84"	1.32

Table 5: Commercial Garage and Aircraft Hangar Doors[1]

Double-Skin Steel Garage and Aircraft Hanger Doors		Open-Piece Tilt-up[a]		Sectional Tilt-up[b]	Aircraft Hangar	
Insulation[a]		**8' x 7'**	**16' x 7'**	**9' x 7'**	**72' x 12' [c]**	**240' x 50' [d]**
1-3/8" thickness	EPS, steel ribs	0.36	0.33	0.34 - 0.39		
	XPS, steel ribs	0.33	0.31	0.31 - 0.36		
2" thickness	EPS, steel ribs	0.31	0.28	0.29 - 0.33		
	XPS, steel ribs	0.29	0.26	0.27 - 0.31		
3" thickness	EPS, steel ribs	0.26	0.23	0.25 - 0.28		
	XPS, steel ribs	0.24	0.21	0.24 - 0.27		
4" thickness	EPS, steel ribs	0.23	0.20	0.23 - 0.25		
	XPS, steel ribs	0.21	0.19	0.21 - 0.24		
6" thickness	EPS, steel ribs	0.20	0.16	0.20 -0.21		
	XPS, steel ribs	0.19	0.15	0.19 - 0.21		
4" thickness	Non-insulated				1.10	1.23
	Expanded polystyrene				0.25	0.16
	Mineral wool, steel ribs				0.25	0.16
	Extruded polystyrene				0.23	0.15
6" thickness	Non-insulated				1.10	1.23
	Expanded polystyrene				0.21	0.13
	Mineral wool, steel ribs				0.23	0.13
	Extruded polystyrene				0.20	0.12
Un-insulated	All products	1.15				

[a]: Values are for thermally broken or thermally unbroken doors.
[b]: Lower values are for thermally broken doors; upper values are for doors with no thermal break.
[c]: Typical size for a small private airplane (single-engine or twin).

[d]: Typical hangar door for a midsize commercial jet airliner.
[e]: EPS is extruded polystyrene, XPS is expanded polystyrene.

Table 6: Doors by Glazing Type [1]

Glass Type		U- Value by Frame Type			
		Aluminum Without Thermal Break	Aluminum with Thermal Break	Reinforced Vinyl/Aluminum-Clad Wood or Vinyl	Wood or Vinyl-Clad Wood/Vinyl Without Reinforcing
Single glazing	glass	1.58	1.51	1.4	1.18
	acrylic/polycarb	1.52	1.45	1.34	1.11
Double glazing	air	1.05	0.89	0.84	0.67
	argon	1.02	0.86	0.8	0.64
Double glazing, e=0.20	air	0.96	0.8	0.75	0.59
	argon	0.91	0.75	0.7	0.54
Double glazing, e=0.10	air	0.94	0.79	0.74	0.58
	argon	0.89	0.73	0.68	0.52
Double glazing, e=0.05	air	0.93	0.78	0.73	0.56
	argon	0.87	0.71	0.66	0.5
Triple glazing	air	0.9	0.7	0.67	0.51
	argon	0.87	0.69	0.64	0.48
Triple glazing, e=0.20	air	0.86	0.68	0.63	0.47
	argon	0.82	0.63	0.59	0.43
Triple glazing, e=0.20 on 2 surfaces	air	0.82	0.64	0.6	0.44
	argon	0.79	0.6	0.56	0.4
Triple glazing, e=0.10 on 2 surfaces	air	0.81	0.62	0.58	0.42
	argon	0.77	0.58	0.54	0.38
Quadruple glazing, e=0.10 on 2 surfaces	air	0.78	0.59	0.55	0.39
	argon	0.74	0.56	0.52	0.36
	krypton	0.7	0.52	0.48	0.32

1. U-factors are applicable to both glass and plastic, flat and domed units, all spacers and gaps.
2. Emissivities shall be less than or equal to the value specified.
3. Gap fill shall be assumed to be air unless there is a minimum of 90% argon or krypton.

As seen in the Tables **1-6**, U-factor of the door varies with a type of door and type of insulation. In addition, U-factor varies with configuration of the door. The reported U-factor also, unfortunately, varies with method of measurement used [2].

ESTIMATION OF PAYBACK PERIOD

When the U-value of a door is estimated using the table or heat sensing equipment, equations from Chapter 2 (windows) may be applied to determine the payback period of replacement of the door. Quick solutions to door efficiency would include placement of insulation at the bottom of the door (in case of garage doors – flaps), caulking, and weather-stripping if needed. The payback for caulking and weather-stripping can be calculated using heat loss equation in Chapter 2. More details on commercial garage doors are presented in Chapter 4.

VENTILATION

On a nice day, businesses chose to keep windows closed and use the door for ventilation purposes. However a study [3] has shown that even at 12 opening per hour, doors may not provide sufficient ventilation that is achieved by having an open window.

CONCLUSIONS

Window replacement is generally a better investment than door replacement from the sustainability standpoint. However, preventing drafts through the door by installing weather-stripping is equally as effective as stopping drafts around windows.

Note: Related assessments found in appendix are: B.

REFERENCES

[1]   Washington State Energy Code 2009 Edition Chapter 51-11 WAC. [Kitsop County website].   Available   from: http://www.kitsapgov.com/dcd/documents/WSEC%2009%20Res%20Amendments.pdf. [Accessed March 2012].
[2]   McGowan A, Jutras R. Testing of Air Leakage and Heat Loss Characteristics of Commercial Door Assemblies. ASHRAE Transactions 2007; 113-2: 414-421.
[3]   Marr D, Mason M, Mosley R, Liu X. The influence of opening windows and doors on the natural ventilation rate of a residential building. HVAC&R Research 2012; 18-1/2: 195-203.

Send Orders of Reprints at reprints@benthamscience.net
Energy Assessments for Industrial Complexes, 2013, 25-27

CHAPTER 4

Garage Doors

Alexander Spivak, Ashok Kumar* and Matthew Franchetti

University of Toledo, Toledo, OH, USA

Abstract: Installation of garage door flaps and other draft prevention methods is generally an inexpensive and easy way to save money. More expensive decisions faced by business owners are insulation of the garage door and interlocking of the door with a heater/air conditioning unit. While former is nearly always good practice, two latter solutions require payback period calculations.

Keywords: Garage doors, garage door insulation, HVAC, cooling garage, heating garage, flap, insulation, energy consumption, interlocking garage door, heat loss, vinyl flaps, workshop garage, garage operations, garage management, garage door opener, furnace, air conditioner, A/C, garage door flaps, temperature controlled garage.

INTRODUCTION

One of the ways to reduce energy consumption is to improve garage doors. There are three major approaches that can be taken either separately or together: installing garage door flaps, improving garage door's insulation, and installing interlocking garage doors [1].

INSTALLATION OF GARAGE DOOR'S FLAPS

Simplest cost saving tip is installation of vinyl flaps for garage door openings. The heat loss (in kilowatts per hour) through the bottom opening of the garage may be estimated by formula:

$$Q = 0.000001117 \times A \times V \times (T_2 - T_1) \tag{1}$$

Where:

  Q is a heat loss

***Address correspondence to Ashok Kumar:** Department of Civil Engineering, The University of Toledo; 3006 Nitschke Hall, Mail Stop 307, 2801 W. Bancroft St., Toledo, OH 43606, USA; Tel: 419-530-8136, Fax: 419-530-8116; E-mail: ashok.kumar@utoledo.edu

A is an area of the opening

V is a wind velocity

T_2 is outside temperature

T_1 is inside temperature

Where area of the opening is length x width (in feet), wind velocity in feet/hour (be conservative: typical good estimate for directional wind speed is 0.0001 ft/hr) and temperature difference between inside and outside is measured in degrees Fahrenheit. Kilowatts per hour can then be multiplied by number of hours and the cost of kw-hr to determine amount of money lost through the opening. Even with a small opening, vinyl flaps can reduce heat loss by as much as 90%, thereby paying for themselves very quickly.

INSULATING GARAGE DOOR

A typical insulated door has an exterior surface, interior surface, an insulating core material, and an "air film, which is due to the insulating effect of air on a vertical surface. Each of these elements has a unique R-value that, added together, create the door section R-Value. R-value is a measure of thermal resistance of material (*e.g.*, doors, insulation).

$$R_{section} = R_{air\ films} + R_{outside\ surface} + R_{insulation} + R_{inside\ surface} \qquad \textbf{(2)}$$

Since some garage doors may not be properly insulated or insulated at all, additional insulation may be considered. Common insulation options are spray-on insulation foam, foam panels and fiberglass panels [2]. Typically, spray-on insulation provides the best effects, since it fills nearly all cavities. Insulation foam is sprayed in the space between inner and outer sheeting of the door [3].

Because each garage door, climate, and use of the garage is different it is difficult to generalize the payback period for any type of insulation for garage doors. Garage sizes, amount of time doors are open, heating systems in the garage, *etc.* significantly affect payback period of the insulation. Your business may want to consider (additional) garage door insulation if:

- The garage or workshop area is heated or cooled.

- If doors are usually closed and furnace or A/C unit is operating on hot or cold days.

Once the decision is executed, the total cost of insulating a garage door may be divided by the subsequent energy savings (due to insulation of garage door) to determine the payback period.

Additionally, the heating/cooling system may be programed so that when the doors are open, effectiveness of heaters may drop by 50–75%. A rule-of thumb estimate would be:

Amount of money saved per year = estimated cost of heating garage × percent of the time heater is on and door is open x 0.5   **(3)**

The simple payback period of interlocking garage door with heater is cost of the unit and installation divided by amount of money saved per year.

CONCLUSIONS

While it is generally a good idea to eliminate drafts from the garage door, the decision to take additional action depends heavily on whether garage is heated or cooled. If separate heating or cooling system is present in the garage, business managers should seriously consider both improving the garage door's insulation (if needed) and linking the garage door with heating and cooling unit. Savings associated with having a linked garage door vary depending on company culture and customer flow.

Note: Related assessments found in appendix are: B, C.

REFERENCES

[1]   Technical Data Sheet #163 from Door & Access Systems Manufacturers Association International   website.   Available   from: http://www.dasma.com/PDF/Publications/TechDataSheets/CommercialResidential/Tds163.pdf . [Accessed March 2012].

[2]   Garage   Door   Insulation   [Garagedooropenerguide   website].   Available   from: http://www.garagedooropenerguide.com/garage-door-insulation.html. [Accessed March 2012].

[3]   Wind Speed Data. [Western Regional Climate Center website]. Available from: http://www.wrcc.dri.edu/htmlfiles/westwind.final.html. [Accessed March 2012].

Send Orders of Reprints at reprints@benthamscience.net

CHAPTER 5

Insulation

Alexander Spivak, Ashok Kumar* and Matthew Franchetti

University of Toledo, Toledo, OH, USA

Abstract: Selection of insulation depends on a large number of factors, such as availability of the space for insulation, climatic conditions, availability of funds and potential savings, *etc.* This chapter discusses different types of insulation available to customers.

Keywords: Insulation, insulation types, insulation selection, blanket batts, blanket rolls, batts and rolls, concrete block, concrete block insulation, foam, foam insulation, foam boards, ISF, ISF insulation, loose fill, loose fill insulation, reflective insulation, spray foam, foam, spray foam insulation SIP, SIP insulation.

INTRODUCTION

There are number of different insulation options available to businesses. They include blanket batts and rolls, concrete block insulation, foam boards, ISFs, loose fill insulation, reflective insulation, spray foams, SIP's, and so on. The selection of insulation may be assisted by the Table **1** presented in this chapter.

INSULATION TYPE

Proper selection of the insulation can make a significant financial difference. However, with such a wide selection of insulating materials available, insulation has become somewhat of an art, as opposed to a precise science. This chapter will introduce reader to various types of insulation and explain insulation selection decisions. Table **1** below describes types of insulation that may be considered:

---

*Address correspondence to Ashok Kumar:** Department of Civil Engineering, The University of Toledo; 3006 Nitschke Hall, Mail Stop 307, 2801 W. Bancroft St., Toledo, OH 43606, USA; Tel: 419-530-8136, Fax: 419-530-8116; E-mail: ashok.kumar@utoledo.edu

Table 1: Insulation types [1]

Form	Insulation Materials	Where Applicable	Typical R-Values Per Inch of Thickness	Installation Method(s)	Advantages	Disadvantages
Blanket: batts and rolls	Fiberglass, Mineral (rock or slag) wool, Plastic fibers, Natural fibers	Unfinished walls, including foundation walls, and floors and ceilings	Standard: 2.9 - 3.8 High performance: 3.7 - 4.7	Fitted between studs, joists, and beams	1) Relatively easy to install 2) Suited for standard stud and joist spacing, which is relatively free from obstructions 3) Optional facing 4) Some fire resistant	1) Hard to fill spaces completely
Concrete block insulation	Foam beads or liquid foam: Polystyrene, Polyisocyanurate or polyiso, Polyurethane, Vermiculite or perlite pellets	Unfinished walls, including foundation walls, for new construction or major renovations	1.1 - 2	Involves masonry skills.	1) Autoclaved aerated concrete and autoclaved cellular concrete masonry units have 10 times the insulating value of conventional concrete	1) Very low insulating ability per inch 2) Not very effective since most of the heat is lost through solid parts of the wall
Foam board or rigid foam	Polystyrene, Polyisocyanurate or polyiso, Polyurethane	Unfinished walls, including foundation walls; floors and ceilings; unvented low-slope roofs.	Polystyrene: 3.8 - 4.4 XEPS: 5.0 Polyiso: 5.6 - 8.0 Polyurethane: 5.6 - 8.0	Interior applications: must be covered with 1/2-inch gypsum board or other building-code approved material for fire safety. Exterior applications: must be covered with weatherproof facing	1) High insulating value for relatively little thickness 2) Can block thermal short circuits when installed continuously over frames or joists 3) Can add structural strength 4) Polyiso makes good roofing insulation	1) Must be specially treated against fire 2) Low sunlight resistant 3) Low moisture resistance 4) Should be treated against insects
Insulating concrete forms	Foam boards or foam blocks	Unfinished walls, including	17.0 +	Installed as part of the building	1) Insulation is literally built into the	1) Expensive

Table 1: contd….

(ICFs)		foundation walls, for new construction		structure.	home's walls, creating high thermal resistance	
Loose-fill	Cellulose, Fiberglass, Mineral (rock or slag) wool	Enclosed existing wall or open new wall cavities; unfinished attic floors; hard-to-reach places	Cellulose: 3.2 - 3.8 Fiberglass: 2.2 - 2.7 Rock wool 3.0 - 3.3	Blown into place using special equipment; sometimes poured in	1) Good for adding insulation to existing finished areas, irregularly shaped areas, and around obstructions	1) Unprofessional installation may cause safety and electrical risks 2) Settles overtime, especially if not properly installed
Reflective system	Foil-faced kraft paper, plastic film, polyethylene bubbles, or cardboard	Unfinished walls, ceilings, and floors	4.0 - 15.0	Foils, films, or papers: fitted between wood-frame studs, joists, and beams	1) Do-it-yourself. All suitable for framing at standard spacing. Bubble-form suitable if framing is irregular or if obstructions are present 2) Most effective at preventing downward heat flow; however, effectiveness depends on spacing	1) Not effective in preventing sideway or upwards heat flow
Rigid fibrous or fiber insulation	Fiberglass, Mineral (rock or slag) wool	Ducts in unconditioned spaces and other places requiring insulation that can withstand high temperatures	4	HVAC contractors fabricate the insulation into ducts either at their shops or at the job sites	1) Can withstand high temperatures	1) Custom installation may be expensive
Sprayed foam and foamed-in-place	Cementitious, Phenolic, Polyisocyanurate, Polyurethane	Enclosed existing wall or open new wall cavities; unfinished attic floors	---	Applied using small spray containers or in larger quantities as a pressure sprayed (foamed-in-place) product	1) Good for adding insulation to existing finished areas, irregularly shaped areas, and around obstructions 2) Can double	1) Very expensive per unit volume

Table 1: contd....

					R-value of each cavity	
Structural insulated panels (SIPs)	Foam board or liquid foam insulation core, Straw core insulation	Unfinished walls, ceilings, floors, and roofs for new construction	polystyrene: 4.0 XEPS: 5.0 Polyiso: 6.0 - 7.0 Polyurethane: 6.0 - 7.0	Builders connect them together to construct a house	1) SIP-built houses provide superior and uniform insulation compared to more traditional construction methods; they also take less time to build	1) Should be treated with insecticide

Above list of insulations is not nearly complete. There is number of different types of insulations currently used as well as being designed. Some examples of such insulations are vacuum insulation and gas-filled panels, aerogels, and phase change materials [2, 3] and ecological thermal insulation, such as fibreboards, loose fill insulation or panels from different plants or their tails [2, 4-6].

In addition to insulation of the building, applying proper insulation to the pipes will have a significant economic effect. This is true even in well-insulated buildings [7]. An optimized insulation can result in significant savings.

WHEN TO CONSIDER INSULATION

Insulation decisions include science as well as some art. The decision to add insulation to the building exterior is often based on manager's gut feeling. (When insulation of the pipe is considered, the decision is typically simpler and may be made by calculating differential temperature (temperature inside the pipe minus temperature of the surface of the pipe)). A decision to insulate exterior (and in some cases interior) walls of the facility may be based on heat loss calculations. If it is not feasible to conduct such calculations, decision may be made based on the following secondary (less scientific) criteria:

- Similar buildings have significantly lower heating/cooling costs.

- Building is old and insulation is either not present or settled.

- Daylight is visible through the cracks/visible deterioration of walls.

- Furnace/A/C unit is properly sized, but working non-stop.

- Furnace/A/C unit is properly sized, but unable to keep steady temperature in the facility.

- Process is temperature sensitive, and additional insulation may help process stability.

There are, of cause, other considerations that may affect insulation decisions. If the decision is made, it is important to monitor savings associated with additional insulation and to compare them to original cost.

INSULATION SELECTION

There is no easy way to decide which insulation is the best. The R-value guidelines for each climatic region are readily available through EPA [8] and many other websites. Recommendations given are generally applicable to residential buildings. Commercial facilities need to decide whether they require more or less insulation based on their business type. For example, a very large facility will have lower heat loss per cubic foot of space per hour than smaller facility. In general, available guidelines are reasonable for most non-manufacturing facilities.

CONCLUSIONS

Typical applications include house and pipe insulation. In general, insulation has relatively short payback period. However, one must estimate payback period prior to purchasing of insulation. Calculations of payback period are impractical in many cases due to their complexity. However, a reasonable estimate can be made and verified over time.

Note: Related assessments found in appendix are: B, C, H, K.

REFERENCES

[1] Energy Star website. Available from: www.energystar.gov. [Accessed March 2012].

[2] Vejeliene J. Processed Straw as Effective Thermal Insulation for Building Envelope Constructions. Eng. Structures and Technologies 2012; 4-3: 96-103.

[3] Jelle BP. Traditional, state-of-the-art and future thermal building insulation materials and solutions – properties, requirements and possibilities. Energy and Building 2011; 43: 2549-2563.

[4] Kymalainen HR, Sjoberg AM. Flax and hemp fibers as raw materials for thermal insulations. Building and Environment 2008; 43: 1261-1269.

[5] Zach J, Korjenic A, Petranic V, Hroudova J, Bednar T. Performance evaluation and research of alternative thermal insulations based on sheep wool. Energy and Building 2012; 49: 246-253.

[6] Pinto J, Paiva A, Varum H, Costa A, Cruz D, Pereira S. Corn's cob as a potential ecological thermal insulation material. Energy and Building 2011; 43: 1985-1990.

[7] Kecebas A. Determination of insulation thickness by means of exergy analysis in pipe insulation. Energy Conversion and Management 2012; 58: 76-83.

[8] Environmental Protection Agency website. Available from: www.epa.gov. [Accessed February 2013].

Send Orders of Reprints at reprints@benthamscience.net

CHAPTER 6

Roofs

Alexander Spivak, Ashok Kumar[*] and Matthew Franchetti

University of Toledo, Toledo, OH, USA

> **Abstract:** Roofing selection depends significantly on purchase price, longevity of the roof, maintenance costs, and energy savings. There are number of roofs that may be considered in each case.

Keywords: Roof types, roof replacement decisions, roof maintenance, roof management, roofing, roof color, roof style, asphalt roof, asphalt roll roof, asphalt roll shingles, concrete roof, fiber reinforced roof, eco roof, eco-roof, ecological roof, rubber roof, plastic roof, membrane roof, burned-on roof, hot mopped roof, hot mopped asphalt roof, decorative roof, metal roof, natural slate roof, slate roof, premium asphalt roof, standard asphalt roof, tile roof, white roof, green roof.

INTRODUCTION

A decision to re-roof a building is typically done when current roof requires extensive repairs. In most cases, it is not economical to replace a roof prior to disrepair in order to decrease cost of utilities, with one exception being a greenhouse roof. However, when re-roofing is required, cost, longevity, maintenance, and utilities savings must be considered. For example, premium asphalt and metal roofs are significantly more energy efficient than standard asphalt or tile roofing. Energy savings vary greatly with local weather. Attic insulation should also be considered as part of energy cost calculations. Roofing company specialists or company engineers can perform the calculations.

SELECTING A NEW ROOF

There are many roofing choices available, including white roofs. Selection of the

---

*Address correspondence to Ashok Kumar: Department of Civil Engineering, The University of Toledo; 3006 Nitschke Hall, Mail Stop 307, 2801 W. Bancroft St., Toledo, OH 43606, USA; Tel: 419-530-8136, Fax: 419-530-8116; E-mail: ashok.kumar@utoledo.edu

new roof should include cost-of-ownership consideration. When deciding on the roof type, payback period may be calculated as difference of cost between the two roofs considered divided by the expected annual savings of the more expensive roof. Table **1** below [1-7] lists the advantages, disadvantages, and various properties of various roofing choices. One must be careful, understanding that roofs that may be good insulators for winter may not necessarily be good insulators for summer. It was suggested by [10] is that one of the most effective means to reduce the additional radiating heat flow from the interior surfaces of the roof coating into the attic during the hot day is to install radiant barriers with low emissivity coefficient into the roof construction.

Table 1: Description of Roof Types [1-7]

	Asphalt Roll Shingles	**Concrete (Fiber Reinforced)**	**Eco-Roofs**
Aestetics/Looks	Average	Excellent	Excellent
Longevity	5 Years	50+ years	50 Years
Storm/Fire proof	Average	Excellent	Excellent
Weathering	Poor	Excellent	Excellent
Energy Savings	Poor	Excellent	Excellent
Green Friendly	Poor	Poor	Excellent
Weight	Light/Medium	Medium	Medium
Noise	Average	Average	Average
Color Fastness	Average	Average	N/A
Color Choices	Poor	Average	N/A
Walkability	Excellent	Excellent	Excellent
Light Susceptibility	Low	Low	Low
Maintenance	Poor	Medium	TBD
Initial Cost	Very Low	High	TBD
Long Term Cost	High	High	TBD
House Style	Flat roofs	Virtually any style of home	Flat to moderately sloped roofs.
Advantages	1) Inexpensive 2) Easy to install and repair	1) Many colors and styles including shakes, tile, and stone 2) Relatively lightweight 3) Fire and insect resistant; meet many of the more restrictive fire codes	1) Environmentally friendly; filters rainwater through a roof system of vegetation and soil 2) Low maintenance; can extend the life of the roof membrane substantially

Table 1: contd….

		4) Low maintenance 5) Extremely durable 6) Resource efficient	3) provides insulation to even out climate variations; in particular, keeps houses cooler in summer 4) Attractive
Disadvantages	1) Very short lived (5 years) 2) Environmentally unfriendly	1) Can be expensive 2) Uneven quality among products	1) Initially expensive 2) Unconventional in US, though used more than 30 years in Europe
Cool Roof Option	White ruberoid		
	Metal	Natural Slate	Premium Asphalt
Aestetics/Looks	Excellent	Exceptional	Exceptional
Longevity	100 years	100+ years	50+ years
Storm/Fire Proof	Excellent	Exceptional	Average
Weathering	Excellent	Excellent	Class A
Energy Savings	Excellent	Poor	Excellent
Green Friendly	Excellent	Excellent	Average
Weight	Exceptionally light	Medium	Light/Medium
Noise	Average	Average	Average
Color Fastness	Excellent	Exceptional	Average
Color Choices	Excellent	Excellent	Excellent
Walkability	Average	Poor	Good
Light Susceptibility	Low	Low	Low
Maintenance	Excellent	Medium	Medium
Initial Cost	High	Very High	Average
Long Term Cost	Very Low	Very Low	Low
House Style	Bungalows, ranch, contemporary, cottage, historic (virtually all)	Colonial, French, Italianate, Exotic Revivals, Chateauesque, Beaux Arts	Can be used on any house from contemporary to historic.
Advantages	1) Available in different looks including cedar shingles, slate, or standing seam 2) Available in copper, aluminum, steel, tin, *etc.* 3) Many colors 4) Light weight 5) Durable 6) Long life span (at least 50 years) 7) Low maintenance 8) Insect and mold resistant 9) Can be	1) Beautiful, distinctive appearance 2) Fireproof 3) Long life span 4) Low maintenance 5) Very durable	1) Many colors, types, thicknesses and manufacturers 2) Suitable for most residential applications 3) Easy to Repair 4) Fire resistant

Table 1: contd….

	installed over existing roofs 10) Excellent performance in high wind, hail and rain 11) Environmentally friendly		
Disadvantages	1) May be difficult to install 2) Can be expensive 3) May need periodic painting	1) Very expensive 2) requires specialized installation 3) Heavy 4) Fragile 5) High maintenance 6) Fails mainly due to fasteners of flashing, not slates themselves	1) Scars easily when hot 2) Environmentally unfriendly
Cool Roof Option	White painted metal		White asphalt shingle with "premium" white granules
	Standard Asphalt	Tile	Wood Shake
Aestetics/Looks	Average	Exceptional	Average
Longevity	15 to 30 Years	100+ years	30+ Years
Storm/Fire proof	Average	Exceptional	Poor
Weathering	Class B	Excellent	Poor
Energy Savings	Poor	Poor	Poor
Green Friendly	Poor	Excellent	Poor
Weight	Light/Medium	Medium	Medium
Noise	Average	Average	Average
Color Fastness	Average	Exceptional	Poor
Color Choices	Average	Excellent	Poor
Walkability	Good	Poor	Good
Light Susceptibility	Low	Low	Low
Maintenance	Poor	Medium	Poor
Initial Cost	Low	Very High	Very High
Long Term Cost	High	Very Low	High
House Style	Can be used on any house from contemporary to historic. False thatched roof with the wrapped roof edge on 1920s Tudor style	Mediterranean, Italian, French Eclectic, Spanish Eclectic, Beaux Arts, Mission, and Prairie. May also be attractive on some contemporary or ranch style homes	Bungalows, ranch, contemporary, cottage, historic
Advantages	1) Inexpensive 2) Ranges from low-cost 3-tab shingle to architectural	1) Non-combustible 2) Many colors and styles 3) Attractive 4) Fireproof	1) Blends in with the environment 2) Easy to repair or replace 3) Long

Table 1: contd….

	shingles with extra durability and style 3) Many colors, types, thicknesses and manufacturers 4) Suitable for most residential applications 5) Easy to repair 6) Can be treated with fire retardants 7) Fire resistant	5) Easy to maintain 6) Very weather -resistant 7) Insect proof, fungus resistant 8) Extremely durable when maintained	lasting if maintained (30–50 years) 4) Can be be pressure treated to resist rot (newer shingles) 5) Can be treated with chemicals to be fire resistant (newer shingles) 6) Composite shingles now available 7) Natural look weathering to a soft grey 8) Offers some insulation value
Disadvantages	1) Relatively short life-span (15–30 years) 2) Scars easily when hot 3) Subject to mildew and moss 4) Environmentally unfriendly	1) Expensive 2) Heavy 3) Used primarily in new buildings because of weight and structural requirements 4) Installation and repairs can be tricky (due to complex fastening) 5) Fragile; walking on roof may break tiles 6) Roof decks must be structurally strong	1) Expensive 2) Usually requires professional installation 3) High maintenance 4) Tends to rot, split, mold, and mildew 5) Poor fire rating unless presure treated
Cool Roof Option	White asphalt shingle	White clay or concrete tile	

	Engineered Rubber/Plastic. Aka Plastic (Membrane) Roofing Aka Burnt-on Roofing	**Hot Mopped Asphalt with Decorative Stone or Gravel (Top Layer)**
Aestetics/Looks	Poor	Poor
Longevity	30+ Years	20 Years
Storm/Fire proof	Poor	Average
Weathering	Average	Poor
Energy Savings	Average	Poor
Green Friendly	Excellent	Poor
Weight	Medium	Medium
Noise	Average	Average
Color Fastness	Excellent	Average
Color Choices	Excellent	Poor
Walkability	Average	Good
Light Susceptibility	Low	Low
Maintenance	High	Medium
Initial Cost	High	High

Table 1: contd….

Long Term Cost	High	High
House Style	Virtually any style of home. More common on commercial structures.	Flat roofed California-style modern, or slightly sloped roofs. Commonly used for commercial buildings
Advantages	1) About 1/3 the weight of slate 2) Long lasting (30–50 years) 3) Cost effective 4) Attractive 5) Available in a large range of styles and colors with more appearing constantly 6) Made of reclaimed materials	1) Inexpensive 2) Easy to repair
Disadvantages	1) New to market	1) May smell bad 2) Health risk to installers 3) Fumes promote smog
Cool Roof Option	White or reflective	Built-up roof with white gravel

MAINTENANCE SUGGESTIONS

Annual maintenance and care may increase the life of an existing roof by as much as 20%. The following should be done regularly:

- Keep branches off the roof.

- Do not be on the roof more than necessary – walking on the roof can cause damages.

- If algae growth on the roof, remove it.

- Check attic for leaks after major storms.

- Keep gutters clean.

SPECIAL CASE: GREEN ROOFS

A lot of attention is given to the research of the green roofs. A "buzz idea" of the green roof does not always materialize in financial or ecological gain. There are articles discussing advantages of common insulation over the green roof, *e.g.,* [8] and application of green roofs in conjunction with other insulation [9]. The latter demonstrated that different types of green roofs perform significantly different, with some being more effective than white roofs.

CONCLUSIONS

When deciding to replace a roof, its total cost of ownership, which includes purchase price, longevity of the roof, maintenance costs, and energy savings, must be considered. Table **1** includes suggestions, such as general long term cost (total cost of ownership). The calculations have to be based on the specific roof.

Note: Related assessments found in appendix are: H, J.

REFERENCES

[1]   Green Roofing Keeping Cool Topside. Available from:
        http://ecomall.com/greenshopping/greenroof.htm [Accessed March 2012].
[2]   Energy Star Website Available from: www.cnergystar.gov. [Accessed March 2012].
[3]   Types of Shingles and Repairs. Available from:
        http://www.hometime.com/howto/projects/roofing/roof_2.htm. [Accessed March 2012].
[4]   Roofing Materials Comparison Chart. Available from:
        http://roofingfundamental.com/Roofing-Material-Comparison.html. [Accessed March 2012].
[5]   Roofing Materials Comparison Chart. Available from:
        http://aluminumshingle.com/comparing_roofing_materials.aspx. [Accessed March 2012].
[6]   Comparison of Roofing Products. Available from:
        http://vancouverislandsbestroof.com/comparison-chart.html. [Accessed March 2012].
[7]   Roofing Materials Comparison Chart. Available from:
        http://www.mackeymetalroofing.com/v/vspfiles/files/education/mmr_roofing_materials_co
        mparison_chart_130510.pdf. [Accessed March 2012].
[8]   Lstiburek JW. Seeing Red Over Green Roofs. ASHRAE Journal 2011; 53-6: 68-71.
[9]   Sailor DJ, Elley TB, Gibson M. Exploring the building energy impacts of green roof design
        decisions – a modeling study of buildings in four distinct climates. Journal of Building
        Physics 2011; 35-4: 372-391.
[10]   Banionis K, Monstvilas E, Stankevicius V, Bliudzius R, Miskinis K. Impact of heat
        reflective coatings on heat flows through the ventilated roof with steel coatings. J. of Civil
        Eng. And Mngmt. 2012; 18-4: 505-511.

Send Orders of Reprints at reprints@benthamscience.net

CHAPTER 7

Appliances

Alexander Spivak, Ashok Kumar[*] and Matthew Franchetti

University of Toledo, Toledo, OH, USA

Abstract: The decision to change appliances is typically made when an appliance is no longer functioning properly, but this chapter challenges that approach. Immediately replacing some appliances may be a sound financial and ecological decision. In addition, we discuss the selection of Energy Star® appliances.

Keywords: Appliances, appliance energy consumption, energy star, appliance replacement decisions, fridge, refrigerator, stove, washer, dryer, washer/dryer combo, stackable, stackable washer/dryer, dishwasher, closes washer, range, self-cleaning range, cost of appliance, life of appliance, smart appliance, smart grid.

INTRODUCTION

In order to optimize costs, business owners are faced with two appliance-related questions: when to replace the appliance, and what to replace it with. The decision is based partially on potential energy savings and partially on the cost of the appliance itself.

WHEN TO REPLACE AN APPLIANCE

Depending on the business, updating an appliance may be one way to save money and make a business more ecologically sustainable. The decision about when to replace an appliance depends on difference in maintenance and energy cost of the existing appliance when compared to a new appliance. The cost of investment into the new appliance would be offset by annual savings.

The payback period of an appliance replacement or removal may be estimated by dividing the total cost of purchase of new unit by the difference in annual energy/

*Address correspondence to Ashok Kumar: Department of Civil Engineering, The University of Toledo; 3006 Nitschke Hall, Mail Stop 307, 2801 W. Bancroft St., Toledo, OH 43606, USA; Tel: 419-530-8136, Fax: 419-530-8116; E-mail: ashok.kumar@utoledo.edu

maintenance cost. The annual energy cost difference is the difference in energy consumption (annual kWh) multiplied by the cost of 1 kWh. Actual energy values for new units are available from the manufacturer typically at the time of purchase. Energy consumption of old appliances may or may not be available, however Table **1** can help estimate energy consumption of appliances if no information about the specific make and model can be found. The simple payback period is a cost of new appliance divided by annual savings (energy and maintenance) due to replacement of the appliance.

Wattages of most appliances are typically stamped on the bottom or back of the appliance or on its nameplate. If the wattage has the range, you may either select an average value or the value of the current setting (if applicable). The wattage can be converted to cost by the following formula:

$$C = W \times H \times 0.052 \times P \tag{1}$$

Where:

  C is annual cost of consumption

  W is wattage

  H is hours of operation per week

  P is cost of kW-hr of electricity

Generally, payback periods of residential appliances range from 5 to 50 years, making some appliances uneconomical to replace while they are functioning properly.

If wattage is not listed on appliance, it may be estimated by the current draw of the appliance. Amperes may be printed on the back of the appliance or on its nametag. If no information about appliance's rating is printed, information may be obtained by hooking an ammeter during the operation cycle, contacting the manufacturer or visiting sites such as www.energystar.gov. Wattage is found by multiplying current by voltage (typically 120V or 240V outlets are used).

The table below present annual energy consumptions for some average-sized appliances sold in Canada. Table **1** may be used for comparison purposes or as a rough estimate.

Table 1: Energy Usage by Appliances [1, 2]

Annual kWh	1984	1990	1997	1999	2004	Energy Star (2011)	2011 Energy Star Savings *vs.* Minimum Standard
Refrigerator	1457	1044	664	664	465	390	10%
Dishwasher	1213	1026	649	640	457	295	10%
Clothes Washer	1243	1218	930	860	573	217	45% elect., 55% water
Standard Clothes Dryer	1214	1103	887	908	912	392	50%
Chest Freezers	813	658	342	337	344	271	10%
Self Cleaning Ranges	790	727	759	742	622	582**	10% elect. or gas

*Data is based on average size appliance. Values provided in the table are guideline values only.
**2009 value

CONSIDERING SMART AND ENERGY STAR® APPLIANCES

The payback period calculations should be made for at least a few prospective replacement candidates. Strong consideration should be given to Energy Star®- rated appliances as they are typically the preferred choice in the long run. Energy Star® appliances are typically about 10% more efficient than newer appliances and often more than 50% more efficient than older appliances. In addition, Energy Star® appliances emit lower levels of greenhouse gases and pollutants than standard appliances.

Smart appliances are devices that can be controlled through some cellular phones, computers or other gadgets. Smart appliances have been designed to allow consumers tighter control over appliances and to effectively shed or reduce domestic usage of energy [3].

CONCLUSIONS

In order to optimize your business from both a financial and a sustainability point of view, replacing appliances may be a good idea. Older appliances, even if they are in fine working order, may be good candidates for replacement if they consume significant amounts of energy. Following the calculation method in the chapter allows business managers to decide whether it is time to replace appliance(s). When making replacement or purchasing decisions, it is a good idea to consider Energy Star$^{®}$- rated appliances as possible candidates.

REFERENCES

[1]   What does it Cost to Operate? [NLCPR website]. Available from http://www.nlcpr.com/ ApplianceData.php. [Accessed March 2012].

[2]   Energy Star website. Available from www.energystar.gov. [Accessed March 2012].

[3]   Grogan, A. Smart Appliances. Eng. and Tech. 2012; 7-6: 44-45.

Send Orders of Reprints at reprints@benthamscience.net

Energy Assessments for Industrial Complexes, 2013, 45-47   **45**

CHAPTER 8

Vending Machines

Alexander Spivak, Ashok Kumar[*] **and Matthew Franchetti**

University of Toledo, Toledo, OH, USA

Abstract: Vending machines usually are not economical to replace, unless the current device's maintenance costs and downtime are high. However, when a decision to replace a machine is made, Energy Star® - rated machines should be considered. The location of the vending machine will also affect overall energy cost, but to a much smaller degree than sales profit.

Keywords: Vending machines, vending machine replacement, vending machine maintenance, vending machine energy consumption, energy use, placing vending machine, vending machine operations, location of vending machine, profit of use of vending machines, vending machine's P&L, why purchase vending machine, cost of vending machine, alternatives to vending machines, Energy Star, vending machine replacement decisions, vending machine lighting, reducing energy cost of existing vending machine, de-lighting vending machine, moving vending machine, where to plug vending machine, management of vending machine.

INTRODUCTION

Vending machines are typically available in facilities either for profit or for the convenience of employees and customers. A company that plans to implement environmentally sustainable solutions may forego vending machines for convenience and encourage employees to bring healthy snacks drinks in reusable containers. Money saving opportunities for vending machines are discussed in this chapter.

REPLACEMENT OF THE VENDING MACHINE

A typical full size vending machine can cost $3,000 to $5,000 with energy costs in the area of $300 per year [1]. Newer machines, equipped with occupancy

---

*Address correspondence to Ashok Kumar: Department of Civil Engineering, The University of Toledo; 3006 Nitschke Hall, Mail Stop 307, 2801 W. Bancroft St., Toledo, OH 43606, USA; Tel: 419-530-8136, Fax: 419-530-8116; E-mail: ashok.kumar@utoledo.edu

sensors, timers and other energy-saving devices can reduce energy usage by 25 to 75% [2], paying for themselves in 20 years. Most businesses' minimum attractive rate of return MARR will not allow for such investment based on energy costs alone. MARR is set by businesses in order to determine whether return on the project is sufficiently high. Therefore, a decision to replace a vending machine must be made for other reasons such as high maintenance costs, significant down time, need for a larger machine, or aesthetics.

When the decision to replace a vending machine is made, Energy Star® machines should be considered. The difference in cost between energy efficient machines and regular machines can be divided by the estimated annual savings in order to determine how quickly the difference in cost will be. Typically such time periods are short, since Energy Star® machines are up to 50% more energy efficient than new machines that does not qualify as Energy Star® units [1].

LOCATING VENDING MACHINE

Location of machine is primarily dictated by proximity to the cafeteria/offices/ shop floor, foot traffic and availability of space/electrical power. When these criteria are considered, a selection of location may also be affected by temperature in the area. Vending machines located in cool and shaded areas tend to last longer, use less energy, and demand less power. Ambient conditions, such as high temperatures or direct solar gain, can make the compressor work harder to maintain the appropriate environment for snacks and beverages [3].

REDUCING COST OF LIGHTING IN THE VENDING MACHINE

Another energy-saving idea is either to disconnect unnecessary lights or to install a timer that turns lights off when not necessary. You should be careful, since disconnecting lights by yourself may lead to damage to the machine. You may ask your vending machine company to de-lamp the advertising lights inside the machine. The lights and ballasts in a typical refrigerated vending machine use about 180 Watts. Some Energy Star® machines are equipped with timers, occupancy sensors and controllers [1].

VENDING MACHINE INSTALLATION AND MAINTENANCE TIPS

Other suggestions of management of vending machines include:

- Make sure everyone involved is aware of and educated about the installation and use of vending machines.

- Coordinate any moves of vending machines. The units use a photo sensor that is permanently mounted to the wall or ceiling over the vending machines.

- Be careful not to overload a circuit where multiple vending machines are plugged into one circuit. Repeaters are available that stagger the starts of multiple machines on one circuit.

CONCLUSIONS

Vending machines, just like many other sources of income, require energy investment. The investment may be balanced with energy-saving ideas in order to optimize the profit. By considering ideas presented in the chapter, managers may save a little money and improve sustainability of the facility.

REFERENCES

[1]   Energy Star website. Available from: http://www.energystar.gov/index.cfm?c=new_specs.vending_machines. [Accessed March 2012].
[2]   Madison Gas and Electric website. Available from: http://www.mge.com/business/saving/madison/pa_50.html. [Accessed March 2012].
[3]   Vending Machines. [North Carolina Department of Environment and Natural Resources website]. Available from: http://www.p2pays.org/energy/Vending.pdf. [Accessed March 2012].

Send Orders of Reprints at reprints@benthamscience.net

CHAPTER 9

Computers and Office Equipment

Alexander Spivak, Ashok Kumar* and Matthew Franchetti

University of Toledo, Toledo, OH, USA

Abstract: A decision to replace computers and office equipment generally should not be based only on electricity demand. Typical decisions may be based on need of specific equipment, or elimination of such need (*e.g.,* fax is no longer needed since company is now using direct fax-to-email system). Most of the savings in the office are associated with reduction or elimination of paper and ink use.

Keywords: Computer, office equipment, EPEAT, energy star, sustainable office management, office equipment energy consumption, office equipment recycling, desktop computer, laptop computer, computer monitor, scanner, copier, laser copier, inkjet copier, fax machine, laser fax machine, inkjet fax machine, multi-function devise, inkjet multi-function devise, inkjet multi-function device, printer, laser printer, inkjet printer.

INTRODUCTION

In order to reduce office equipment cost, a manager may consider eliminating equipment, establishing policies to shut off equipment when not in use (overnight, weekends) and replacing equipment with Energy Star products when upgrades are needed. EPEAT® rating may also be used in decision-making process. EPEAT® is an environmental rating system for computers and other electronic equipment.

ELIMINATION OF OFFICE EQUIPMENT

Depending on application, some files may not require paper storage at all. Files may be backed up twice and encoded twice, nearly eliminating chances of file loss, and significantly reducing chances of hacking. Often backup is completed

*Address correspondence to Ashok Kumar: Department of Civil Engineering, The University of Toledo; 3006 Nitschke Hall, Mail Stop 307, 2801 W. Bancroft St., Toledo, OH 43606, USA; Tel: 419-530-8136, Fax: 419-530-8116; E-mail: ashok.kumar@utoledo.edu

off-site by data storage/backup vendors. Since it is a common practice for enterprises to protect and keep electronic versions of files anyway, a direct cost saving of electronic storage of files is:

$$S = P + I + R + C + F + M + E - B \tag{1}$$

Where:

S is savings

P is cost of paper

I is cost of ink

R is cost of recycling paper

C is cost of cartridges

F is cost of filing papers

M is cost of securing confidential materials

E is cost of additional electricity

B is cost of electronic data storage and backup

Avoiding on number of papers unnecessarily printed in the facility, it may be possible to save thousands of dollars per month.

ENERGY SAVING OPTIONS

Savings by using Energy Star® office equipment vary depending on the product. However, such savings are fairly small, and payback periods are long. Greater savings may be achieved by turning office equipment off at night. This practice saves utility cost, reduces wear on equipment and helps to protect computers from viruses and cyber attacks. Table **1** below is composed on the basis of EPA data.

Table 1: Electricity Costs of Office Equipment [1]

	Annual Electricity Cost Savings with Energy Star Equipment (Computers are on / not in Sleep Mode)	Annual Electricity Cost Savings with Energy Star Equipment (Computers are on / in Sleep Mode at Night)	Annual Electricity Cost Savings with Energy Star Equipment (Computers are off at Night)
Desktop Computer	$21	$3	$2
Laptop Computer	$6	$1	$1
Computer Monitor	$6	$1	$1
Scanner	$0.27	$0.27	$0.27
Copier			
Laser - Monochrome	$8	$8	$8
Laser - Color	$4	$4	$4
Fax Machine			
Ink Jet	$1	$1	$1
Laser	$8	$8	$8
Multifunction Device			
Ink Jet	$2	$2	$2
Laser - Monochrome	$8	$8	$8
Laser - Color	$24	$24	$24
Printer			
Ink Jet	$1	$1	$1
Laser - Monochrome	$3	$3	$3
Laser - Color	$17	$17	$17

The data in the Table **1** shows average devices. Detailed data for new devices is often available through manufacturer. The best conclusions from this data table are:

- Turn office equipment off at night.

- Include these estimated savings into total cost of ownership calculations when considering new office equipment.

To obtain good estimates on savings, an EPA website www.energystar.gov can be used.

Selection of equipment plays a significant role in energy consumption. A study conducted by [2] concluded that the energy use of office equipment in Japanese offices with no use of power management is lower than that in US offices with maximum use of power management because portable computers are much more popular in Japan, and the manual-off rate at night is two times higher than in the US. [2] also concluded that power management is very effective for reducing the energy use of office equipment.

RECYCLING OF OLD OFFICE EQUIPMENT

When office equipment completes its useful life, a viable alternative to disposal would be to find a new use. Computer recycling programs exist and are typically operate as nonprofit organizations (NPOs) Through which unwanted office equipment may be donated to schools both in US and overseas or offered to the firm's employees (possibly through employee exchange website). Alternatively, office equipment is accepted by de-manufacturing firms, which typically use parolees and work-released prisoners to disassemble. It is a good idea to know what kind of certification these firms have. NAID-certified firms must properly destroy all data, while R2 and E-Stuart certified firms must properly manage electronic waste. Disassembly operations of American and European office equipment in third world countries exist, but are not known to provide reasonably safe working conditions for their employees and are generally not recommended for partners.

CONCLUSIONS

When it comes to office equipment, the most financially sound policies would be to eliminate faxes if possible, and reduce or eliminate the use of paper and ink (especially inkjet printers). Replacement of the equipment is dictated by both needs of employees and maintenance demands. The industry standard of a four year life span on computers may not be applicable to many businesses. Turning off equipment when it is not in use is an easy, sustainable, and economical practice.

Note: Related assessments found in appendix are: B, F, N.

REFERENCES

[1]   Energy Star website. Available from: www.energystar.gov. [Accessed March 2012].
[2]   Kawamoto K, Shimoda Y, Mizuno M. Energy saving potential of office equipment power management. Energy and Building 2004; 36: 915-923.

Send Orders of Reprints at reprints@benthamscience.net

Energy Assessments for Industrial Complexes, 2013, 52-53

CHAPTER 10

Showers

Alexander Spivak, Ashok Kumar[*] and Matthew Franchetti

University of Toledo, Toledo, OH, USA

Abstract: Proper selection of shower heads and proper management practices in shower rooms may result in significant water use savings with little investment.

Keywords: Shower, commercial shower, shower head, shower head flow, water consumption, shower head replacement, shower head selection, water usage, water cost, water management, low-flow shower head, showering techniques, proper showering techniques, use of shower facilities, management of shower facilities, types of shower head, water consumption, shower head replacement cost, showering facility, commercial showering facility, managing cost of water.

INTRODUCTION

Reducing the overall water flow in showers can be achieved by two methods: technological and sociological. The technological method entails replacing existing shower heads with low-flow shower heads.

SHOWER HEAD REPLACEMENT

An older shower head can process as much as 6.5 gallons per minute, while low-flow shower head disperses 2.2 to 2.5 gallons per minute [1]. The savings in the commercial shower room allow for an average savings of $5.40 per shower per person (estimating 0.15 cents per gallon of water and 8-10 minutes average shower time).

MANAGEMENT APPROACH

In addition to showerhead replacement, management may place signs with proper

---

*Address correspondence to Ashok Kumar: Department of Civil Engineering, The University of Toledo; 3006 Nitschke Hall, Mail Stop 307, 2801 W. Bancroft St., Toledo, OH 43606, USA; Tel: 419-530-8136, Fax: 419-530-8116; E-mail: ashok.kumar@utoledo.edu

showering techniques (*e.g.,* turning water off while using soap) and/or hire temporary "example providers" – people whose job would be to follow instructions while taking showers.

COMMERCIAL SHOWER SYSTEMS

Commercial shower systems water saving techniques may also include water-filtering process. The closed-loop system allows for the recycling and re-use of water in a similar manner, as it is done in the pool. The decision to install such system depends on the amount of use of showering facilities, since such system has a significant installation cost as well as maintenance cost. The payback period of such system is expected to be long.

CONCLUSION

By following the above recommendations, facilities that use showers may significantly reduce their costs in a very quick payback period.

REFERENCE

[1]   Green 3D Home website. Available from: http://www.green3dhome.com/YourHouse /Bathroom/Showershowerhead.aspx. [Accessed March 2012].

Send Orders of Reprints at reprints@benthamscience.net

CHAPTER 11

Pools

Alexander Spivak, Ashok Kumar* and Matthew Franchetti

University of Toledo, Toledo, OH, USA

Abstract: Covering the pool and using the correct pool heater are the most cost-effective and sustainable ways of managing the outdoor pool.

Keywords: Commercial pools, pool, outdoor pool, pool heating systems, pool covers, pool flow rate, pool pump, pool management, pool covering, pool cover type, water management, water, pool heater type, pool heater replacement, pool pump downsizing, outdoor pool, pool pipes, pool pipes length, pool pipes diameter, electric pool heater, gas pool heater, solar powered pool heater.

INTRODUCTION

There are number of ways to reduce the cost of heating an outdoor pool. These methods include covering the pool, using an economical heater, and reducing length and diameter of the pipes.

BEST PRACTICES

If an outdoor pool is part of your business, using a pool cover is the best sustainability advice available, reducing total energy cost of running a heat pump by 84% ± 9% [1]. Another big source of savings comes from the type of pool heater used. Heat pump pool heaters are approximately twice as economical as gas pool heaters (depending of gas pool heater's efficiency) and five times more economical than electric pool heaters. Or, as an alternative, solar pool heaters represent one of the most cost-effective uses of solar energy. Depending on collector type, installation design and location of the business, solar panels

*Address correspondence to Ashok Kumar: Department of Civil Engineering, The University of Toledo; 3006 Nitschke Hall, Mail Stop 307, 2801 W. Bancroft St., Toledo, OH 43606, USA; Tel: 419-530-8136, Fax: 419-530-8116; E-mail: ashok.kumar@utoledo.edu

typically range between 60 and 110% of the area of the pool. In addition, reducing the time of use of the pump and substituting the filter for a larger one (and keeping intake clean) may further reduce energy use.

OTHER SUSTAINABLE COST-SAVING OPPORTUNITIES

Downsizing an existing pump to the required minimum decreases the flow rate and, therefore, both energy consumption and maintenance costs. In addition increasing diameters of existing pipes, decreasing angles of flows and/or shortening pipes allow water flow to be more smooth, reducing energy use of the pump.

The recommended turnover rate of commercial pools is six hours which makes the desired flow rate (DFR) of the pumps as the number of gallons in the pool divided by turnover rate and multiplied by 60. The turnover rate is calculated by the number of times the pump circulates the entire pool volume through the filter system in 24 hr. period. Typically, commercial pools turn over every 6 hrs., while residential 8 to 10 hrs. Exceeding the DRF results in direct energy and machine wear-related losses. DRF is also affected by presence of fountains, spas, waterfalls, solar heating, and in-floor cleaning systems among other things, if they operate through the same pumping system.

Increasing the diameter of the pipes, decreasing lengths of pipes, and reducing the number of sharp angles (such as replacing 90° angle turns with a set of 45° angle turns) decreases total dynamic head (TDH) of the pump, therefore, decreasing energy consumption. Total dynamic heat in feet of water is the amount of pressure resistance in the pool piping. It is approximately equal to pressure at the filter (in psi) multiplied by 2.31 [1]. The pressure can be read with a simple inexpensive meter. Some pumps have a built in pressure readout.

CONCLUSIONS

Suggestions presented in the chapter can help significantly reduce the cost of the heating pools as well as equipment wear. Covering the pool is a very effective way of saving costs with little investment. In addition, gas heaters or solar panels instead of electric heater will usually pay back within a short period of time. Other improvements, such as improving pumps and pipes, may also improve the performance and reduce costs of running a pool. Return on investment calculations depends on the differences between current and proposed systems. Simple payback period is calculated as cost of investment divided by annual reduction in energy and maintenance costs.

REFERENCE

[1]   Energy Savers website. Available from: http://www.energysavers.gov/your_home/water_heating/index.cfm/mytopic=13290. [Accessed March 2012].

Send Orders of Reprints at reprints@benthamscience.net
Energy Assessments for Industrial Complexes, 2013, 57-64

CHAPTER 12

Air Flow of Kitchen Hoods

Alexander Spivak, Ashok Kumar[*] and Matthew Franchetti

University of Toledo, Toledo, OH, USA

Abstract: The two main methods of improving kitchen hood air flow are air flow adjustment (low investment, high return) and system replacement (high investment, high return). There are a number of other ways to improve the efficiency of the kitchen hoods.

Keywords: Kitchen hood, kitchen hood maintenance, kitchen hood air balance, kitchen hood air flow, kitchen hood replacement, types of kitchen hoods, overhead canopy kitchen hood, backshelf kitchen hood, proximity kitchen hood, wall hood, wall kitchen hood, single island hood, single island kitchen hood, double island hood, double island kitchen hood, energy saving hood, energy saving kitchen hood, short cycle hood, short cycle kitchen hood, continuous mode hood, continuous mode kitchen hood, economizer mode hood, economizer mode kitchen hood, exhaust hood, exhaust kitchen hood, tempered air hood, tempered air kitchen hood.

INTRODUCTION

The kitchen hood is a major component of kitchen ventilation system. It is a major source of energy consumption, which is taking its energy through HVAC. There are number of different types of kitchen hoods, each with its advantages and disadvantages. Understanding options may help businesses to select the best system available on the market.

TYPES OF KITCHEN HOODS AND THEIR PROPERTIES

Kitchen Hood types include overhead canopies and backshelf/proximity hoods. The latter are typically used when overhead canopies are impractical. Overhead

*Address correspondence to Ashok Kumar:** Department of Civil Engineering, The University of Toledo; 3006 Nitschke Hall, Mail Stop 307, 2801 W. Bancroft St., Toledo, OH 43606, USA; Tel: 419-530-8136, Fax: 419-530-8116; E-mail: ashok.kumar@utoledo.edu

canopies are further subdivided into wall hoods (installed against the wall), single island hood (installed over a single island) and double island hoods (installed as island over double line of equipment).

Each of these three types of hoods are further subdivided into energy saving hoods (short cycle hoods available in both continuous and economizer modes), exhaust hoods and tempered air hood [1, 2]

Energy saving (short cycle) and its economizer version hoods intake filtered un-tempered outside air. By reducing airflow, they can save energy (especially in econo-mode). However, they are generally not used over the open flame cooking equipment [1-4]. Tempered air hoods can take tempered or un-tempered air [1-4]. Exhaust hoods do not introduce air. They are used for grease exhaust applications or moisture/heat capture [1-4]

SUSTAINABLE SAVINGS OPPORTUNITIES

There are many opportunities for energy savings with kitchen hoods including airflow adjustment and replacement of the system. Other opportunities are presented at the end of the chapter.

AIR BALANCE ADJUSTMENT

Air balance is adjusted using the premise air in = air out. If not enough air is coming in, the system will work harder to try to squeeze air out, possibly resulting in high energy use, grease contamination and fires. The common symptom of poor air balance is hard-to-open doors [5].

Air balance calculations entails following steps [1-5]:

$$F = L \times W \times 50 \tag{1}$$

Where:

F is base air flow

L is hood length (in feet)

W is hood width (in feet)

Equipment adjustment is shown in Table **1**:

Table 1: Exhaust Air Requirement Per Equipment Type [1, 2, 5]

Equipment Type	Examples	Additional Exhaust Air Needed
Low Temperature (400°F)	ovens roasters steam kettles ranges non-grease producing equipment	0 ft^3/min of air per each ft^2 of appliance's surface
Medium to High Temperature (600°F)	fryers grilles griddles deep fat fryers	35-40 ft^3/min of air per each ft^2 of appliance's surface
Maximum Temperature (700°F)	solid fuel appliances char broilers	90-100 ft^3/min of air per each ft^2 of appliance's surface

For the "heat only" hoods (moisture and hood removal), following equipment adjustments are needed are shown in Table **2a**:

Table 2a: Exhaust Air Requirement Per Load Type [2, 5]

Equipment Type	Examples	Additional Exhaust Air Needed
Low Load	non-grease producing equipment such as ovens, rotisseries	40 -65ft^3/min of air must be removed per 1 ft^2 of hood opening
High Load	dishwashers	80-100 ft^3/min of air must be removed per 1 ft^2 of hood opening

For exhaust and proximity hoods, exhaust air quantities may be estimated as shown in Table **2b**:

Table 2b: Exhaust Air Requirement Per Load Type [2, 5]

Equipment Type	Examples	Additional Exhaust Air Needed
Low Load	steam equipment kettles	200 +/- ft^3/min of air must be removed per 1 ft of hood length

Table 2b: contd...

Medium	ranges, ovens griddles, grilles	300 +/- ft³/min of air must be removed per 1 ft of hood length
High Load	fryers	500 +/- ft³/min of air must be removed per 1 ft of hood length

Please be aware that estimates above are for guideline purposes only.

To compare duct velocity to the rating of the fan, following equation is used:

$$Duct\ velocity(ft/\min) = \frac{air\ flow(ft^3/\min)}{duct\ area(ft^2)}$$

Calculations of the pressure drop are more involved. These calculations entail determination of filter pressure drop and exhaust collar losses. If you have pressure drop data, it may be compared to fan's rating.

REPLACEMENT OF THE SYSTEM

When considering system replacement, it is advantageous to consider high efficiency systems, as they have shorter relative payback periods. The calculation of the payback period is done by dividing total cost of the system by the expected annual energy savings. The selection of the proper kitchen hood is not a straight forward process. In fact, firms responsible for kitchen hood installation do not always sell best suited product. For example, a study [6] concluded that among studied 89 hoods in 60 restaurants, 39% of the hoods met flow rate guidelines of the American Conference of Governmental Industrial Hygienists and 24 % met the guidelines of the American Society of Heating, Refrigerating and Air-Conditioning Engineers. In less than 4% of the cases, inadequate flow rates were identified.

The selection of system is further complicated by the position of appliances as well as range top type and accessories. Even subtle changes in appliance position and hood configuration could dramatically affect the exhaust rates required for complete capture and containment, regardless of the appliance duty and/or usage [7]. There is also a significant relationship between appliance usage and accessories, as well as hood dimensions (including overhang, reservoir volume, and mounting height) and hood C&C (capture and containment) performance [8].

OTHER SAVING OPPORTUNITIES AND SUGGESTIONS

There are a number of general guidelines and recommendations for hood optimization. Following recommendations below may save significant percentage of the total energy and maintenance costs of the hood, making your business more sustainable and profitable.

Considerations for energy savings, as recommended by experts [1-5]:

- Check for damage to interior angles close to or at the edge. Damage will significantly reduce capture and containment (C&C) performance of the hood.

- Check the feasibility of installing a variable speed fan. Such installation may significantly decrease energy consumption. They can adjust motors speeds down to 10% capacity, which generates substantial savings (around 40%) with payback periods of 1 to 5 years. Some systems claim as much as 50% energy reduction.

- When a kitchen is not in use, exhaust fans should be shut off or significantly reduced.

- NFPA recently reduced the minimum duct velocity to 500 fpm at the exhaust collar of the ductwork. Typical fans operate at three times the speed. It may be possible to make adjustments that will reduce the energy cost and still satisfy the guidelines. Adjustments are best made if variable speed system is available or may be installed.

- Make adjustments to air flow as appliances change. See required air flow per each type of appliance above. Review your cooking equipment positioning in reference to the hood. Push your equipment as far back as possible. It may be advantageous to place heavy-duty appliance at the center of the hood station.

- Install side and/or back panels on the canopy to reduce heat gains, increase effectiveness, reduce the effect of cross-drafts and improve capture rates.

- When possible, introduce transfer air (such as air from the dining room) that has already been conditioned to the right temperature.

- Avoid 4-way or slot ceiling diffusers in the kitchen, especially near hoods. They introduce cross-drafts and may cause discomfort to kitchen personnel and reduce C&C rates. Also, consider other techniques to keep make up air speed (especially near hood) below 75 fpm.

- Seal the gap behind equipment to improve efficiency.

- Calculate feasibility of integrating the CKV system with the total HVAC system to create the most effective, efficient system.

- Consider new high efficiency filters. While they are not energy savings, some filters, like disposable wool or reusable metal filters, may reduce the frequency of hood cleanings.

- Consider a kitchen air balance check if you suspect a problem (for example, if doors hard to open).

- Install a demand ventilation system to reduce the total air flow for exhaust and make-up air.

- Install high efficiency motors for your fans.

- Maintain good operating conditions of your hood system (cleaning, obstructions, belts, bearings, *etc.*). Keep filters clean/replace in time to avoid pressure loss increase that will force the fan to work harder

- Consider testing for plume spillage (may be expensive).

- If possible, consider reducing air temperature. Capture and containment performance of hoods is significantly better at 75°F then at 90°F.

- Reduce total exhaust quantities and total make-up air quantities (see methods above).

- Consider human component of kitchen operations. It may account for a large portion of the energy bill.

- If your kitchen exhaust fan is running at higher pressure than needed, you may be losing money. Make sure that fan properly covers all areas.

CONCLUSIONS

Kitchen hood systems are reasonably complicated. It is recommended that major decisions are made with help of the facility engineer. The guidelines presented in the chapter can be used to reduce the overall cost of air and solve common problems.

Note: Related assessments found in appendix are: A, C, H, J.

REFERENCES

[1]   Food Service Technology Center website. Available: http://www.fishnick.com/ventilation. [Accessed March 2012].
[2]   Kees Hood and Fan Selection and Application Guidelines. Available: http://www.kees.com/pdf/agkh.pdf. [Accessed March 2012].
[3]   Improving Commercial Kitchen Ventilation System Performance. Available:s www.fishnick.com and www.archenergy.com. [Both sites accessed March 2012].
[4]   Improving Energy Efficiency in Commercial Kitchens. Available: http://www.energyright.com/business/pdf/Kitchens_SC_ESCD.pdf. [Accessed March 2012].
[5]   Energy Conscious Blueprint: Commercial Kitchen Variable Volume Exhaust Hood. Available: www.cl-p.com and www.uinet.com. [Both sites accessed March 2012].
[6]   Keil C, Kassa H, Fent K. Kitchen Hood Performance in Food Service Operations J. of Env. Health 2004; 12: 25-30.

[7]   Swierczyna R, Soboiski P, Fisher D. Effects of Appliance Diversity and Position on Commercial Kitchen Hood Performance. ASHRAE Transactions 2006; 112-1: 591-602.

[8]   Soboiski P, Swierczyna R, Fisher D. Effects of Range Top Diversity, Range Accessories and Hood Dimensions on Commercial Kitchen Hood Performance. ASHRAE Transactions 2006; 112-1: 603-612.

Send Orders of Reprints at reprints@benthamscience.net
Energy Assessments for Industrial Complexes, 2013, 65-69   65

CHAPTER 13

Motors

Alexander Spivak, Ashok Kumar[*] and Matthew Franchetti

University of Toledo, Toledo, OH, USA

Abstract: Installing or reprogramming Variable Speed Drives (VSD) / Variable Frequency Drives (VFD) may be a very effective way to reduce energy consumption of motors. The selection and programming of the VSD are best for the professionals. When the device is properly programmed, quick payback periods are typically expected.

Keywords: Motors, motor maintenance, VSD, motor efficiency, VSD programming, variable speed drive, VFD, VFD programming, variable frequency drive, motor selection, motor replacement, VSD maintenance, VSD installation, VFD maintenance, VFD installation, variable speed drive maintenance, variable speed drive installation, variable frequency drive maintenance, variable frequency drive installation, mechanical VFD, hydraulic VFD, electrical VFD.

INTRODUCTION

If variable loads are exerted upon a motor-driven piece of equipment, a variable frequency drive (VFD) or variable speed drive (VSD) can make a significant difference in the amount of energy used. A VFD varies the speed of a connected motor by changing the voltage supplied to the motor. It allows continuous processing of the speed control by converting from fixed frequency voltage to the continuous variable frequency. Depending on applications of VFD, motor speed is matched to the load. Properly programmed VFDs may save up to 40% of energy costs.

TYPES OF VFDS

There are several different types of VFDs including Volts/Hz, Voltage Vector, Voltage Vector +, Flux Vector and Servo.

*Address correspondence to Ashok Kumar: Department of Civil Engineering, The University of Toledo; 3006 Nitschke Hall, Mail Stop 307, 2801 W. Bancroft St., Toledo, OH 43606, USA; Tel: 419-530-8136, Fax: 419-530-8116; E-mail: ashok.kumar@utoledo.edu

Volts/Hz (Basic Scalar) drives, are the least expensive drive with the fewest features. This drive is usually setup for CT and Open Loop. Voltage/Space Vector drives can be set for CT/VT and Open/Closed Loop. Voltage Vector Plus drives can be set for CT/VT and Open/Closed Loop. In addition, they can calculate motor characteristics without spinning the shaft of the motor. Flux Vector drives can perform all functions of Voltage Vector Plus drives with greater level of accuracy and dynamic responsiveness. Some Flux vector drives require special motors. Servo drives have even more superior dynamic responsiveness characteristics. However they are significantly more expensive and do not operate with AC induction motors.

Mechanical VSDs are more commonly used than others. Yet this seems to change, as cost of other more sophisticated options reduce compared to energy savings. Mechanical variable speed control usually uses a belt drive which is controlled by moving conical pulleys manually or with positioning motors [6-8] Hydraulic VFDs work on the principle of oil volume differential. There are two types of hydraulic VSDs including hydraulic pump and motor, and fluid coupling [6, 8, 9]. The electrical VSDs using power converter controls.

ADVANTAGES OF VFDS

There are number of reasons to use a VFD [1-4]:

- To match speed to changing load requirements (improves efficiency).

- To allow accurate and continuous process control over a wide range of speeds.

- To control process parameters such as temperature, pressure or flow without the use of a separate controller.

- To reduce maintenance costs (lower operating speeds result in longer life for bearings and motors).

- To eliminate the throttling valves and dampers when possible.

- To eliminate a need for soft starter for the motor [1, 2].

- To have controlled ramp-up speed (in a liquid system this can eliminate water hammer problems).

- To limit torque to a user-selected level (can protect driven equipment that cannot tolerate excessive torque).

- To reduce current peaks when starting large motors with VFD that gradually ramp the motor up to speed.

- To manage input power based on system demand and use only the energy required by the driven equipment [2, 3].

- To regenerate power which can be routed back to the system or sold to utilities for additional revenue (some VFDs) [4].

A review article [6] concluded that VSDs are reliable and cost effective means to control the speed of electrical motors. Installing VSDs on electrical motor applications improves the efficiency of the systems and saves energy. They require little maintenance, provide the most energy efficient capacity control, have the lowest starting current of any starter type, and reduce thermal and mechanical stresses on motors and belts. In addition, they protect the motor while keeping the process running, reduce pump failure caused by pump cavitations, and reduce maintenance on piping and valves. Applying VSDs to the HVAC systems and compressed air reduces energy consumption.

WHEN TO CONSIDER VFDS

VFDs save energy in a wide range of, but are most effective in machines that operate significantly below maximum speed for extended periods of time such as blowers, boilers, centrifugal pumps, compressors and fans. If motors operate at constant speed, VFDs are less economical. In fact, many experts consider VFDs unnecessary in constant speed operations. VFD drives may be optimized using various techniques, including trial and error. Since sensors that can record voltage consumption of the motor are available and may be connected to most motors, it is possible to measure actual savings due to the addition of a VFD.

Prior to making a decision about the installation of a VFD it is important to check the compatibility and effectiveness of the device with a motor and to understand the benefits and costs of installation. Simple payback calculations may involve multiple components, such as the cost of VFD and its installation, energy savings, repairs, maintenance and downtime estimates (both of VFD and the equipment) and change in longevity of the equipment as result of installation of the VFD. The simple payback period varies from six months to five years, but is typically under a year.

Consult the manufacturer or facility engineer for detailed knowledge of VFD equipment operation and process requirements to ensure energy savings. Generally, manufacturers will have VFD-selection software available to customers.

VFD's may be used for a single motor or multiple motors. However, if a single VFD is used with multiple motors (less expensive arrangement), individual overload protection, such as thermal overloads are needed to protect each motor.

REWIRING MOTORS

While motors appear to be functional, they may have extremely low power factors in one or two legs of a three-phase power, which can result in energy loss and may require special attention. Low power factors may be due to incorrect power at the motor, but are more likely due to a reversal of "DELTA" and "WYE" wiring configurations in the distribution system. If Delta and Wye are reversed, engineer may choose to either rewire the motor (price: cost of rewiring + 1% efficiency loss + downtime) or to replace it with a higher efficiency motor.

ALTERNATIVE FUELS

As prices for conventional fuels increase, alternative options are now considered in large motor applications. Alternative fuels used include cornstalk, straw, sweet sorghum (used for bio-ethanol, bio-gasoline and bio-diesel), sugar beat bagasse (bio-ethanol and bio-gasoline), and pork fat (bio-diesel) [5]

PAYBACK PERIOD ESTIMATION

Payback period for the installation of VFD drives on most motors is short, and can be estimated by dividing the total cost of installation of the device by annual energy savings. VFD ratings help engineers estimate the payback period.

CONCLUSIONS

When practical (for example, with blowers, boilers, centrifugal pumps, compressors and fans), a VFD may be considered as viable economic and sustainable option. Often installation and proper programming of VFD drives as well as re-programming existing ones can significantly cut energy costs. Since cost of the devices is relatively low compared to cost of energy, VFDs should be seriously considered for many types of motors.

Note: Related assessments found in appendix are: J, M, O, P, R, S, U.

REFERENCES

[1] National Resources of Canada website. Available from: http://oee.nrcan.gc.ca/industrial/equipment/vfd/savings.cfm?attr=24. [Accessed March 2012].

[2] Danfoss Group Global website. Available from: www.danfoss.com. [Accessed March 2012].

[3] Energy Savings with Variable Speed Drives [Rockwell Automotive website]. Available from: http://samplecode.rockwellautomation.com/idc/groups/literature/documents/ar/7000-ar002_-en-p.pdf. [Accessed March 2012].

[4] St. Patricks of Texas website. Available from: http://www.stpats.com/index.htm. [Accessed March 2012].

[5] Adnadjevic BK. New technologies in the production of motor fuels from renewable materials. Thermal Science 2012; 16-1: S87-S95.

[6] Saidur R, Mekhilef S, Ali MB, Safari A, Mohammed HA. Applications of variable speed drive (VSD) in electrical motors energy savings. Renewable and Sustainable Energy Reviews 2012; 16: 543-550.

[7] Barnes M. Practical variable speed drives and power electronics. Technology and Engineering, Australia 2003.

[8] ABB. Guide to variable speed drives [ABB website]. Available from: www05.abb.com [Accessed December 2012].

[9] Okaeme N. Automated robust control system design for variable speed drives. Dissertation 2008, Univ. of Nottingham, UK.

Send Orders of Reprints at reprints@benthamscience.net

CHAPTER 14

Boilers

Alexander Spivak, Ashok Kumar[*] and Matthew Franchetti

University of Toledo, Toledo, OH, USA

Abstract: There are number of different sustainable ways to reduce the cost of boiler use with good paybacks. Methods of boiler cost reduction are presented in this chapter. The key concept of this chapter is that cost of boiler is significantly less than cost of operation of the boiler.

Keywords: Boilers, boiler maintenance, boiler efficiency, boiler blowdown, boiler economizer, boiler heat loss, fire tube boiler, water tube boiler, longitudinal drum boiler, cross drum boiler, bent tube boiler, stirling boiler, packaged boiler, fluidized bed combustion boiler, atmospheric fluidized bed combustion boiler, pressurized fluidized bed combustion boiler, circulating fluidized bed combustion boiler, stoker-fired boiler, spreader stoker boiler, chain grate stoker boiler, traveling grate stoker boiler, pulverized fuel boiler, waste heat boiler, tubeless boiler, condensing boiler.

INTRODUCTION

The most efficient way to generate steam and to transfer its energy to the plant is through the use of boilers. Currently many types of boilers are available for use. This chapter will discuss replacement of boilers as well as improvements in modern boiler efficiencies.

Two most common types of boilers are shell boilers and water tube boilers. Shell boilers (wet and dry back configurations) include Lancashire boilers, economic boilers (with two or three passes of tubes from burner to chimney), boilers with four passes, packaged boilers and reverse flame/thimble boilers. In water tube

---

*Address correspondence to Ashok Kumar: Department of Civil Engineering, The University of Toledo; 3006 Nitschke Hall, Mail Stop 307, 2801 W. Bancroft St., Toledo, OH 43606, USA; Tel: 419-530-8136, Fax: 419-530-8116; E-mail: ashok.kumar@utoledo.edu

boilers, unlike shell boilers, water is circulated inside the tubes, with the heat source surrounding them. Water tube boilers include longitudinal drum boiler, cross drum boiler, bent tube and Stirling boiler. Because they are more complicated to operate, water tube boilers are generally used when required operating pressure exceeds 400 psi.

Table **1** [1-7] provides general information about various types of boilers.

Table 1: General Boiler Information

Common Boiler Type	Description of Operations
Fire tube boiler	Hot gases pass through the tubes and boiler feed water in the shell side is converted into steam between 2 and 4 times. The turn around zones can be either dryback (refractory-lined) or water-back (water-cooled).
Water tube boiler: layout types: longitudinal drum, cross drum, bent tube (stirling) boilers	Feed water flows through the tubes and enters the boiler drum. The circulated water is heated by the combustion gases and converted into steam at the vapour space in the drum
Packaged boiler	Generally of shell type with fire tube or water tube design, pre-packaged and almost ready for operations as delivered. Can be 2, 3, or 4-pass
Fluidized Bed Combustion Boiler (variations include: Atmospheric Fluidized Bed Combustion Boiler, Pressurized Fluidized Bed Combustion Boiler, Circulating Fluidized Bed Combustion Boiler)	The bed of solid particles exhibits the properties of a boiling liquid and assumes the appearance of a fluid when air velocity is sufficiently high– "bubbling fluidized bed".
Stoker Fired Boiler (includes: Spreader Stokers, Chain-grate or Traveling-grate Stoker)	In spreader stockers coal is continually fed into the furnace above a burning bed of coal. The coal fines are burned in suspension; the larger particles fall to the grate, where they are burned in a thin, fast-burning coal bed. In chain grate stockers coal is fed onto one end of a moving steel grate. As grate moves along the length of the furnace, the coal burns before dropping off at the end as ash.
Pulverized Fuel Boiler	The pulverized coal is blown with part of the combustion air into the boiler plant through a series of burner nozzles.
Waste Heat Boiler	Used in the heat recovery from exhaust gases from gas turbines and diesel engines. May need additional sources of energy
Tubeless and Condensing Boilers	"Tubeless" Boilers use tubing coils instead of rigid tubes. "Direct Contact" water heaters have no tubes, tubing or coils; they have heat transfer media such as spheres or cylinders and allow flue gases to come in direct contact with the water.

Table 1: contd….

Common Boiler Type	Typical Steam Capacities*	Typical Steam Pressures**	Typical Fuels***	Efficiency Range
Fire tube boiler	small, up to 27,000 lb/hour	low to medium, up to 250 psi	oil, gas, solid fuels	Increases with number of passes.
Water tube boiler: layout types: longitudinal drum, cross drum, bent tube (stirling) boilers	high, 9,000 to 270,000 lb/hour	very high, up to 3000 psi	oil, gas, solid fuels	Can be very efficient by themselves, can reach 90% if combined with small waste heat boilers
Packaged boiler	varies greatly, 8,000 to 260,000 lb/hour	varies greatly, 250 to 3000 psi	oil, gas, solid fuels	Increases with number of passes. Four-pass can be very efficient
Fluidized Bed Combustion Boiler (variations include: Atmospheric Fluidized Bed Combustion Boiler, Pressurized Fluidized Bed Combustion Boiler, Circulating Fluidized Bed Combustion Boiler)	have a wide capacity range- 1,000 to over 220,000 lb/hour	50 to 1,200 psi	coal, washery rejects, rice husk, bagasse & other agricultural waste	FBC boilers can operate with overall efficiency of 84% (±2%). CFBC boilers are generally claimed to be more economical than AFBC boilersfor industrial application requiring more than 75 - 100 T/hr of steam
Stoker Fired Boiler (includes: Spreader Stokers, Chain-grate or Traveling-grate Stoker)	have a wide capacity range- 200 to over 220,000 lb/hour	50 to 1,250 psi	gas, solid fuels	generally lower (most 70-85%), two controlling factors of efficiency from the combustion system are excess air and carbon loss
Pulverized Fuel Boiler	up to 11,000,000 lb/hour	up to 9,000 bar	solid fuels	most 70 -90%
Waste Heat Boiler	usually small	50 to 2,500 psi	water heat, oil, gas, solid fuels	most 60 -90%
Tubeless and Condensing Boilers	usually small or medium	50 to 2,000 psi	oil, gas, solid fuels	up to 97%

* Steam capacities are shown as saturated steam.
** Steam pressures are based on 0 psig, 212F.
*** Fuels used in boilers are coal (anthracite, bituminous, brown, lignite, peat, semi bituminous), oil (diesel or gas oil, light, medium, or heavy fuel oil), gas (natural gas, LPG – liquefied petroleum gas), wood, electricity or solid wastes. Multiple fuels may also power boilers. Other fuel modifications are possible.

BOILER'S EFFICIENCY

The efficiency of a boiler is a measure of its ability to generate the steam demand from a given fuel supply [9, 10]. Without condensation, newer boilers' efficiencies range from 74% to 85% [2]. Boilers that are designed for condensation and use advanced controls to squeeze every possible BTU from the combustion process are able to achieve efficiencies up to 97% [2]. Use of heat recovery devices in the flue gas path brings efficiency factors up to 98–99%.

Boiler selection is based on a number of factors. They were well described [9, 11] to include operation purpose, manufactured steam amount, pressure and temperature, the inlet temperature of feed water, water hardness, fuel type, lower heating value of fuel, fuel analysis, and fuel cost.

One important factor to consider as part of boiler's purchase is efficiency. Efficiency ratings are available from manufacturers. however, a purchaser must be careful not to compare apples to oranges when it comes to efficiency. Sometimes the representation of boiler efficiency does not truly represent the comparison of energy input and energy output of the equipment [7].

Main sources of inefficiency are stack loss and tank heating. After these two sources are addressed, engineer should start looking for other sources. A neglected unit may be as much as 50% inefficient. A crude estimate for a "medium" size commercial boiler would be 1% increase in efficiency corresponds to $1,000 in annual fuel savings, making purchase price of the boiler small compared to the fuel cost. This estimate is, of course, only a rule of thumb, not a guideline.

BOILER'S EFFICIENCY CALCULATION

Combustion ratio analysis is an important step in boiler efficiency management. Fuel requires a set amount of air in order to burn completely. If the amount of air too great, the exccss air will cause reduction in burning temperature/loss through the stack, and, subsequently, boiler's efficiency. However, some access air is necessary in order to burn fuel completely. Insufficient oxidation of fuel will result in planar pollution as well as lower temperatures. Again, efficiency of the boiler will suffer. Smaller boilers typically use "open loop" control of fuel-to-air mixture. Such system requires periodic recalibration or operator control. Larger plants utilize computerized/operator controls in "closed loop" system. Currently, more systems are moving to automatic controls, guided by sensory data fed into computer program.

Efficiency of the boiler is defined as a ratio of heat:

$$Boiler\ efficiency\ (\eta) = \frac{Heat\ exported\ in\ steam}{Heat\ provided\ by\ the\ fuel} = \frac{Q(h_g - h_f)}{q(GCV)} \times 100\%$$ **(1)**

Where:

Q is quantity of steam generated per hour (in kg/hr)

h_g is enthalpy of saturated steam in kcal/kg of steam (available from charts as function of temperature and gauge pressure)

h_f is enthalpy of feed water in kcal/kg of water (available from charts as function of temperature and gauge pressure)

q is quantity of fuel used per hour (in kg/hr)

GCV is gross calorific value of the fuel (in kcal/kg of fuel)

Boiler efficiency may be improved by introduction of aftermarket controllers. They may reduce energy consumption up to 11% for a 24 hr period and reduce boiler cycling by up to 57% [12]. It is also recommended to perform a complete PM overhaul of the boiler on the annual basis – preferably during scheduled plant shutdown [13].

SOURCES OF INEFFICIENCIES IN BOILERS [1-3]

The principle losses that occur in a boiler are:

-   Moisture in fuel.

-   Moisture in combustion air.

-   Excess air (newer firing gaseous and liquid fuels operate at excess air Levels of 15%, older systems can have over 25% excess air).

-   Dry flue gas.

- Flue dampers.

- Blowdown (steam boilers).

- Unburned fly ash (newer systems will have nearly 0% excess fuel).

- Unburned bottom ash (newer systems will have nearly 0% excess fuel).

- Radiation (possibly due to cladding micro-damage, insulation).

- Convection.

- Loading (due to boiler sizing, see multiple boilers information further in the chapter).

- Missing/damaged economizer.

- Heat recovery system.

- Concentrate return.

- Electrical systems (see Chapter 8).

- Pipelines – excessive footage of pipes, damage to pipes, insufficient insulation (see Chapter 5).

- Temperature of feed water (can pre-heater improve bottom line? Generally for every 6°C increase in feed water fuel savings increase by 1%) [7].

HEAT LOSS DUE TO FLUE GASES

Heat loss in the flue gases is the biggest single source of reducible heat loss. Increase in stack temperature results in direct loss of efficiency of the boiler. Newer technologies can recover energy from the flue gas, practically eliminating this problem. However, without such a system, engineers may still reduce the stack heat losses. Engineers may easily measure heat loss at the stack, check heat transfer surfaces for contamination, check burner(s) calibration or maintenance and check condition and position of dampers. A sensor present at the stack area is critical for such operations. It is a good idea to keep some type of alarm that indicates both high and low temperatures of the stack. Lower temperatures (below due point) can cause physical damage to the unit. Boilers with reverse fan and radial damper design that operate properly may also show higher efficiencies.

Each boiler's operation stack temperature is designed to operate within a specific range (typically varying from 200°F to 400°F, depending on the system) [4]. The stack heat may be collected and used either as part of pre-heating system or diverted to another part of the plant. Since the excess heat is essentially free, payback depends on cost of piping, insulation, maintenance and initial cost of heat. Excess heat may also be collected from other areas of the plant in a similar fashion.

HEAT LOSS DUE TO LOAD

Efficiency of the boiler is also affected by the load. The maximum efficiency of the boiler occurs at about two-thirds of the full load [4]. In general, efficiency of the boiler reduces significantly below 25% of the rated load and operation of boilers below this level should be avoided as much as possible [4].

ECONOMICS OF BLOWDOWN

Boiler blowdown eliminates solid deposits that result in corrosion. Depending on the system, blowdown may be intermittent (bottom blowdown) or continuous. Continuous blowdown is generally more desirable option [3]. However, uncontrolled continuous blowdown is very wasteful [3]. Automatic blowdown controls can be installed that sense and respond to boiler water conductivity and pH. A 10% blow down in a 15 kg/cm^2 boiler results in 3% efficiency loss [4]. In addition, automatic blowdown may work in conjunction with water heat recovery system, increasing efficiency further [4].

Heating of the feedtank will reduce the amount of sodium sulfite ($Na_2(SO_3)$) added to the boiler feedwater. This will reduce the amount of bottom blowdown needed or frequency of automatic blowdowns [4]. However, this method results in additional heating costs that must be accounted for.

Estimate for the payback for the heat recovery from the blowdown may be calculated by determining amount of energy collected (quantity of flash stream) by measuring temperature, pressure, and using steam tables [3, 4].

$$Percent\ flash\ steam = \frac{Change\ in\ entropy\ of\ water}{Entropy\ of\ steam} = \frac{h_{f,high\ pressure} - h_{f,low\ pressure}}{h_{g,low\ pressure}} \times 100\% \quad (2)$$

The maximum amount of total dissolved solids (TDS) concentration permissible in boiler is typically provided by manufacturers. TDS levels range from 2,000 ppm in low pressure water tube boilers up to 10,000 ppm in Lancashire boilers [11]. Permissible amount of TDS in specific boiler generally should not be exceeded because damage, downtime and efficiency costs will exceed costs of blowdown.

Other impurities in water are shown in Table **2**.

Table 2: Impurities in Water [5-8]

Name	Symbol	Common Name	Effect
Calcium Carbonate	$CaCO_3$	caulk, limestone	soft scale, carbonic acid
Calcium Bicarbonate	$Ca(HCO_3)_2$		soft scale + CO_2, carbonic acid
Calcium Sulphate	$CaSO_4$	gypsum	hard scale
Calcium Chloride	$CaCl_2$		Corrosion
Hydrogen Sulfide	H_2S		Corrosion
Magnesium Carbonate	$MgCO_3$	limestone	soft scale, carbonic acid
Magnesium Sulphate	$MgSO_4$	magnesite	Corrosion
Magnesium Bicarbonate	$Mg(HCO_3)_2$	epsom salts	scale, corrosion
Sodium Chloride	$NaCl$	common salt	Electrolysis
Sodium Carbonate	Na_2CO_3	soda	Alkalinity
Sodium Bicarbonate	$NaHCO_3$	baking soda	priming, foaming
Sodium Hydroxide	$NaOH$	caustic soda	alkalinity, embrittlement
Sodium Sulphate	Na_2SO_2	glauber salts	Alkalinity
Silicon Dioxide	SiO_2	silica	hard scale

ECONOMIZERS

Boiler's economizer is a device for warming feed water with gases entering the chimney or stack. Installation of economizers is one of the effective ways to improve efficiency. Newer boilers will typically come with economizers and older boilers can be retrofitted with an economizer. Stack Economizers should be considered when large amounts of make-up water are used or needed [2, 5, 6, 8]. The simple payback of the use of economizer can be calculated by accounting for retrofitting/purchasing costs, change in downtime, change in water and energy consumption. There are number of different types of economizers. The selection of economizer depends on boiler size and configuration.

REPLACEMENT ANALYSIS OF BOILERS

Boilers with multi-fuel burning capacities (including unconventional fuels, such as old mattresses) are becoming more popular due to rising and unpredictable cost of fuels. Some boiler designs that meet these criteria are supercritical boilers (designed to operate above steam's critical pressure of 3208 psi) with advance PFBC technology and integrated coal gasification combined cycle [8].

A-PFBC system uses heat of the flue gas as a heat source for partial gasifier. IGCC uses a combined cycle format with a gas turbine driven by the combusted syngas, while the exhaust gases are heat exchanged with water/ steam to generate superheated steam to drive a steam turbine. Using IGCC, more of the power comes from the gas turbine. Typically 60–70% of the power comes from the gas turbine with IGCC, compared with about 20% using PFBC [7, 8].

A decision to replace boiler is typically a capital investment. In order to make such decision, facility must consider following aspects [7, 8]:

- Costs of replacement of the boiler (total purchase price of the new boiler less profit/loss from sale of existing boiler, removal of current boiler, installation of new boiler, downtime, learning curve, *etc.*).

- Maintenance costs of defender and challenger boilers.

- Total downtime costs of defender and challenger boilers.

- Difference in energy costs of defender and challenger boilers.

- Difference in environmental and legal costs of boilers (including consideration of any upcoming legislation).

-    Future production plans.

The payback period may be estimated by calculating annual savings for each year of operations (adjusted by company's MARR – if required) and adding values until total investment cost is reached As always, it is a good practice to pre-calculate values for a few candidate boilers that meet output, fuel and environmental constraints of the facility.

Good replacement candidates would include boilers requiring overhaul, older inefficient boilers (older Primitive, Autogate, *etc.* boilers), under-sized boilers (oversized as well, if there is no future need for increased amount of steam), and boilers with high fuel costs.

When multiple boilers are required, a centralized boiler plant may allow for cost reduction due to standardization of parts, increase in fuel selection, reduction in downtime and management costs. Well controlled staged boilers will reduce unnecessary cycling and energy consumption, as well as need for repairs and downtime.

CONCLUSIONS

Boilers often offer significant sustainable cost savings, some obvious (such as good maintenance practices) and others not so noticeable. When making decisions with boilers it is important to remember that cost of the machine and maintenance is often significantly smaller than cost of the steam generated. An engineer can best determine how to optimize boilers.

Note: Related assessments found in appendix are: A, D, E, I, L, T.

REFERENCES

[1]   The National Gas Boiler Burner Consortium website. Available from: http://cleanboiler.org/Eff_Improve/Primer/Boiler_Introduction.asp. [Accessed March 2012].
[2]   Energy Efficiency Guide for Industry in Asia website. Available from: www.energyefficiencyasia.org/energyequipment/typesofboiler.html. [Accessed March 2012].
[3]   Technical Study Report on Biomass Fired Fluidized Bed Combustion Boiler Technology for Cogeneration from UN Environmental Programme Division of Technology Industry

and Economics website. Available from:
http://www.unep.fr/energy/activities/cpee/pdf/FBC_30_sep_2007.pdf. [Accessed March 2012].

[4] Johnson N, Fundamentals of Stoker Fired Boiler Design and Operation [presented at CIBO Emission Controls Technology Conference July 15-17, 2002 by from Council of Industrial Boiler Owners Website]. Available from: http://www.cibo.org/emissions/2002/a1.pdf. [Accessed March 2012].

[5] Boiler Technologies: Existing and Emerging Trends [Energy Managers Training website]. Available from:
http://www.energymanagertraining.com/Journal/Boiler%20Technologies.pdf. [Accessed March 2012].

[6] Cleaver Brooks website. Available from: www.boilerspec.com. [Accessed March 2012].

[7] SAIE website. Available from: www.saie.com. [Accessed January 2012].

[8] The National Certification Examination for Energy Managers and Energy Auditors website. Available from: http://www.em-ea.org/Guide%20Books/book-2/2.2%20Boilers.pdf. [Accessed March 2012].

[9] Kaya D, Eyidogan M. Energy Conservation Opportunities in an Industrial Boiler System. J. of Energy Engineering 2012; 136-1: 18-25.

[10] Kilicaslan I, Ozdemir E. Energy economy with a variable speed drive in an oxygen trim controlled boiler house. Int. J. of Energy Resources 2005; 127: 59-65.

[11] Onat K, Genceli O, Arisoy A. Heat calculation in steam boiler, Istambul 1988.

[12] Rowley P, Glanville P. Laboratory Evaluation of Aftermarket Boiler Control System. ASHRAE Transactions 2012; 118-2: 230-236.

[13] Bulley, D. Boiler Life-Cycle Considerations. Heating / Piping / Air Conditioning Engineering 2012; 84-9: 42-49.

Send Orders of Reprints at reprints@benthamscience.net
Energy Assessments for Industrial Complexes, 2013, 81-83

CHAPTER 15

Plant Floor Machinery

Alexander Spivak, Ashok Kumar[*] and Matthew Franchetti

University of Toledo, Toledo, OH, USA

Abstract: Engineers typically make plant machinery's economic replacement decisions. Such decisions involve cost analysis, which allow determining when the current machine is to be replaced with a new one. A manager's job may be to suggest when such analysis should be conducted, especially on older machines, and when production changes occur.

Keywords: Floor plant machinery, machine replacement decisions, machine maintenance, replacement analysis, MARR, minimum acceptable rate of return, reliability, accessibility, cost of repairs, cost of maintenance, downtime, machine retention cost, machine replacement cost, forecasting, cost forecasting.

INTRODUCTION

Machinery replacement decisions are significant part of economic business practices that may save large sums of money for a facility.

MACHINE REPLACEMENT

Engineers typically make economic decisions about the plant machinery based on a number of factors, such as machine's throughput, energy consumption, downtime (reliability and accessibility), cost of repairs and maintenance, and financial considerations such as total initial investment, initial downtime and learning curve, company's MARR (minimum acceptable rate of return) availability of funds and affect of investment on company's effective tax rate. From the sustainability standpoint, a machine may become a candidate for replacement (formal term: defendant) if similar new machines are significantly

---

*Address correspondence to Ashok Kumar: Department of Civil Engineering, The University of Toledo; 3006 Nitschke Hall, Mail Stop 307, 2801 W. Bancroft St., Toledo, OH 43606, USA; Tel: 419-530-8136, Fax: 419-530-8116; E-mail: ashok.kumar@utoledo.edu

more economical. More economical machines may either use less energy or use existing and currently wasted sources of energy.

The machine replacement decision includes interest and tax calculations. However, an estimated decision can be made without including these factors. The decision will then include calculation of the cost of retaining current machine for a period of time *vs.* cost purchase of replacement machine and keeping it for the same period of time:

$$C = (L + E) \times H + M + O \pm P \qquad\qquad (1)$$

Where:

C is Cost of keeping current machine

L is Labor cost per hour

E is Energy cost per hour

H is Hours of operation needed to meet estimated customer demand over the period of time

M is Cost of maintenance and repairs

O is Cost of overtime associated with downtime

P is Profit from sale of the machine at the end of the period or total disposal costs

$$C = T + N \pm Po + (L + E) \times H + M + O \pm P_N \qquad\qquad (2)$$

Where:

C is Cost of keeping current machine

T is Total cost of purchase/installation of a new machine

N is Cost of learning to use new machine

P_O is Profit from sale/trade-in the old machine or old machine's disposal costs

L is Labor cost per hour

E is Energy cost per hour

H is Hours of operation needed to meet estimated customer demand over the period of time

M is Cost of maintenance and repairs

O is Cost of overtime associated with downtime

P_N is Profit from sale of the machine at the end of the period or new machine's total disposal costs

Some of the above values have to be estimated (forecasted) based on available information. The longer the period studied in this analysis, the greater is the estimation error and the greater the need to consider interest rates. In addition, a replacement decision must consider potential major changes in technology, company's ability to compete, need to meet quality requirements of customers, and work environment.

Machines that are not in use may be turned off. It is typically an engineer's job to set up a list of SOPs that determine when specific machines or entire lines may be turned off or placed in energy-saving mode. Decisions are based on ramp-up costs and time and needs and production and maintenance needs. It is not uncommon for machines to be left on because no one is assigned to turn them off. General practices on the plant floor are easy to observe but may not be as easy to correct.

CONCLUSIONS

Replacement analysis of floor machinery that includes interest rate and tax calculations is outside of the scope of this eBook. However, such information may be found in many eBooks on engineering economics.

Note: Related assessments found in appendix are: R.

Send Orders of Reprints at reprints@benthamscience.net

CHAPTER 16

Compressors

Alexander Spivak, Ashok Kumar[*] and Matthew Franchetti

University of Toledo, Toledo, OH, USA

Abstract: Compressors may offer opportunities for cost savings. Four major areas of such are system pressure reduction, installation of system controls, installation of variable speed drive (VSD), and leak repairs. Each of these areas requires attention of an engineer.

Keywords: Compressors, compressor maintenance, multiple compressors, compressor controls, VFD, VSD, compressor leaks, positive displacement compressors, reciprocating compressors, single-acting reciprocating compressors, double-acting reciprocating compressors, diaphragm reciprocating compressors, rotary compressors, rotary compressors with helical screw, rotary compressors with liquid ring, scroll rotary compressors, sliding vane rotary compressors, lobe rotary compressors, dynamic compressors, centrifugal dynamic compressors, axial dynamic compressors.

INTRODUCTION

Compressors, along with boilers and motors have great potential for energy reduction. Compressors can consume around 20% of the total electricity bill of the factory [3]. The US Department of Energy (2003) reports that 70 to 90% of the compressed air is lost in the form of unusable heat, friction, misuse, and noise [3]. The cost of the compressed air over time becomes much greater than purchase and installation cost of compressor. Therefore, replacements and overhauls of compressors often have short payback periods. Energy savings generated by replacement or repairs to system, increased productivity and reduced downtime will typically result in significant cost avoidances for the company.

*Address correspondence to Ashok Kumar: Department of Civil Engineering, The University of Toledo; 3006 Nitschke Hall, Mail Stop 307, 2801 W. Bancroft St., Toledo, OH 43606, USA; Tel: 419-530-8136, Fax: 419-530-8116; E-mail: ashok.kumar@utoledo.edu

TYPES OF COMPRESSORS

There are a number of different types of compressors. They include:

- Positive displacement compressors;

  o Reciprocating compressors:

    • Single-acting reciprocating compressors.

    • Double-acting reciprocating compressors.

    • Diaphragm reciprocating compressors.

  o Rotary compressors:

    • Rotary compressors with helical screw.

    • Rotary compressors with liquid ring.

    • Scroll rotary compressors.

    • Sliding vane rotary compressors.

    • Lobe rotary compressors (straight or helical).

- Dynamic compressors;

  o Centrifugal dynamic compressors.

  o Axial dynamic compressors.

  o Combined staging dynamic compressors.

Compressor types are compared in the Table **1** below [1-8]:

These types of compressors are further subdivided by the number of compression stages, drive type, lubrication type, cooling medium, and customization details.

Table 1: Efficiency by Compressor Type

Item	Reciprocating Compressor	Rotary Vane Compressor	Rotary Screw Compressor	Centrifugal Compressor
Basic characteristics	Increase the pressure of the air by reducing its volume (positive displacement machines) Single stage machine's pressure at 70-100 psig Two-stage machine's pressures at 100-250 psig	Increase the pressure of the air by reducing its volume (positive displacement machines) Single stage machine's pressure at 200+/- psi	Increase the pressure of the air by reducing its volume (positive displacement machines) Most typical are single stage helical or spiral lobe oil flooded screw compressors Also, oil free compressors available	Depends on transfer of energy from a rotating impeller to air (dynamic displacement compressor) Oil free by design Surge limit rating at 60%-65%
Efficiency at full load	75%-85%	70%-75%	75%-85%	70%-80%
Efficiency at no load (power as % of full load)	10%-25%	30%-40%	25%-60%	20%-30%
RPM Range	300-1,000	400-1,500	Many at 1,500	3,000-40,000
Flow Range, ACFM	10-3,000	10-3,000	10-20,000	100-200,000
Noise level	Quiet to Noisy	Quiet	Quiet-if enclosed	Quiet
Size	Varies	Compact	Compact	Compact
Oil carry over	Moderate	Low-Medium	Low	None
Vibration	Low to High	Almost none	Almost none	Almost none
Maintenance	Many wearing parts	Few wearing parts	Very few wearing parts	Sensitive to dust in air
Capacity	Low-High	Low-Medium	Low-High	Medium-High
Pressure	Medium-Very High	Low-Medium	Medium-High	Medium-High

MAJOR CHARACTERISTICS OF COMPRESSORS

Supply side of the compressed air system includes air treatment system and compressors. Demand side includes distribution and storage systems and end-use equipment. Major components of the compressed air system include: intake air filters, inter-stage coolers, after-coolers, air-dryers, moisture drain traps, receivers, piping network, filters, regulators and lubricators.

There are number of ways to reduce the overall cost of compressed air. These methods apply to all compressors and compressor systems. First, an age of compressor significantly affects its performance. Even well maintained units degrade over time, reducing its efficiency. As technology changes, the efficiency gap between older and newer units increase even further. It is a generally good idea to consider newer compressors.

Control of air intake temperature may be important. Air temperature pre-treatment may be considered in extreme climates. A generally accepted rule is percent of relative air delivery decreases 1/3% and power savings decreases 1/4% per 1°C increase in inlet air temperature [4].

Pressure sensors around the filter are good indicators of a potential efficiency drop of the compressor. The pressure differential alarm would indicate a need for replacement of the filter. If no sensors are available, regular filter replacement is important. Initially, for 100 mm WC of the pressure drop power consumption of compressor increases by 0.75% to 1.5% [4]. If problem is still neglected, power consumption of the compressor will increase further at faster pace and may cause mechanical breakdown.

SOURCES OF COST REDUCTION

There are four major sustainability source reductions for compressors: system pressure reduction, installation of system controls, installation of variable speed drive (VSD), and leak repairs.

SYSTEM PRESSURE REDUCTION

Reducing the system pressure may generate significant power savings (up to 10%, typically 1% for every 2 lb/in^2 reduction in discharge pressure [3, 4]). However, system pressure optimization has to be done carefully to avoid not having enough pressure for the needs of the facility. If needs of the facility are not constant (more common case), multiple types of software are available to optimize pressure, and therefore, save energy. Consult plant engineer on adverse effects of use of software.

SYSTEM CONTROLS

VSD Drives

Variable Speed Drives are commonly used with compressor system, since load changes in compressor systems are common. A properly installed VSD will have a relatively short payback period. Selection of type of VSD is a prerogative of plant's engineer. Savings associated with use of VSD include leak reduction, lower discharge pressure, stability of operation and reduction in wear. A superior driver option to achieve a variable speed operation is the steam turbine. Gas turbine can also offer variable speed, but some heavy frame gas turbines mandate relatively limited speed ranges [8].

Leak Controls

Air leaks is a common problem that increase with wear. Maintenance of compressors require periodic search for leaks. Leaks may be responsible for up to 30% of compressor's output [3]. Typically, leaks occur at connections, joints, couplings, hoses, tubes, fittings, pressure regulators, open condensate traps, shut off valves, disconnects and thread sealants. A worn out compressor valve, for example, can reduce the compressor capacity by as much as 20% [3]. Leaks can be detected using conventional method, such as soap and water. However, the more effective method would be ultrasonic acoustic detectors.

Other Sustainable Maintenance Practices

There are many other maintenance issues and practices that affect sustainability as well as bottom line. They include [1-8]:

- Monitor air flow delivery sensor. Deviation in the air flow delivery by more than 8% usually requires corrective action. A "nozzle test" is the easiest way to check air flow.

- Proper cooling methods may avoid losses of up to 5% of specific power. Also, very low cooling water temperature will cause moisture in the air and can damage cylinder.

- Maintain clean conditions and proper temperature of intercoolers and after-coolers.

- It is very typical for facilities to have an oversized compressor. The reason is that facilities often purchase compressor to meet future expansion and/or worst case scenario demands. There is an energy cost associated with oversized compressors. To identify whether compressor is oversized, consider off load *vs.* load running hours ratio. Consider replacement analysis or introduction of VSD to oversized compressors. Also, consider multiple compressors (or system) if facility demand varies significantly. It is important to remember that cost of compressor is only a fraction of the total cost of compressed air.

- Properly size inlet air piping. A pressure drop gradient of 1+% is an indicator of inefficiency.

- Tee connections that feed air into a moving stream of air at a right angle result in excess turbulence and pressure drop. Directional tees can reduce pressure drop by 3 to 5 lb/in^2.

- Dead-head connections that feed two streams of air directly at each other result in excess turbulence and pressure drop. Directional connections can reduce pressure drop by 10 lb/in^2.

- Compressed air piping should be sized based on peak flow rate and pipe length. The outlet pipe of many air compressors has a 2-inch diameter. This is rarely enough to minimize pressure loss through the distribution system.

- Review diameters of the hoses used. An undersized hose may result in significant pressure drop.

- "Quick disconnects" may be a significant source of leaks.

- Consider restricting use of the compressed air for low-pressure applications. The cost of compressed air is significantly higher than use of the low pressure air. It may be advantageous to make low pressure air available where needed. Investment a few blowers may have a very quick payback.

- Check oil pressure daily.

- Check filters at least monthly. Consider sensors that may indicate pressure drop across filters. The inlet air filter is most susceptible to deposits. Depending on environment, it may need to be checked very regularly.

- Many systems have condensate traps to gather and (for those traps fitted with a float operated valve) flush condensate from the system. Manual traps should be periodically opened and re-closed to drain any accumulated fluid; automatic traps should be checked to verify they are not leaking compressed air.

- In refrigerated dryers, inspect and replace pre-filters regularly as these dryers often have small internal passages that can become plugged with contaminants.

- Regenerative dryers require an effective oil-removal filter on their inlets, as they will not function well if lubricating oil from the compressor coats the desiccant. The temperature of deliquescent dryers should be kept below 100°F to avoid increased consumption of the desiccant material, which should be replenished every 3-4 months depending on the rate of depletion.

- Floating head pressure reduces the compressor related energy required as well as providing other system advantages and reduces the overall energy used if condenser fans and other equipment is not operated excessively to compensate for an excessively low head pressure.

- Periodically check for leaks and pressure losses throughout the system.

- Avoid cracking drains to achieve moisture free performance at a particular point-of- use.

- Maintain all point-of-use operations at the lowest possible pressure.

- Don't use of air hoists and air motors.

- When not in use, shut off air supply.

- Isolate single users of high pressure air.

- Monitor pressure drops in piping systems.

- Do not use modulating compressors.

- Carefully select types of motors used. Replacement of a motor with one with higher efficiency may have short payback period.

- Consider multiple staged compressors (be careful with this consideration, it may end up costing more).

- Lower the output pressure as far as possible.

- Use waste heat off the compressor to help the rest of the plant save energy.

- Deliver high pressure only to areas of need.

- Understand multiple compressor system controls is very important in optimization of the process.

- Utilize intermediate controls/expanders/high quality back pressure regulators.

- Understand the requirements for clean- up equipment.

- Consider air drying technology with maximum allowable pressure dew point.

- Use cool outside air for the compressor intake. Cooling and even pre-filtering intake air may be practical, but not always.

- Adopt a systematic preventive maintenance strategy for your compressor.

- Pressure drop may be due thin, long pipelines, corroded pipelines or improper connectors.

- Consider training programs.

- Maintain good housekeeping.

- Ensure that condensation can be removed swiftly quickly from the distribution network, or does not occur.

- Check that receivers are sized to store air for short heavy demands.

- Maintain peak compressed air system performance by managing entire system.

- Consider not to skimp on piping when making up equalizer lines. Oversized piping is preferred to undersized piping. General practice indicates the use of oil equalizer lines equal to the full size of the tapping in the compressor.

- If oil is used, select proper oil.

OPERATING MULTIPLE COMPRESSORS

Where more than one compressor feeds a common header, compressors must be operated in such a way that the cost of compressed air generation is minimal. It is easy to see if compressors are nearly full loads: they are very hot to touch.

Following suggestions help in multi-compressor systems [6]:

- When similar compressors are used, set all but one to full load, and let remaining compressor handle the balance of the load (modulate).

- When different size compressors are used, let the smallest unit to handle the balance of the load.

- When different size compressors are used together, let compressor with lowest no load power to handle the balance of the load.

- Consider power consumptions of compressors to decide which one should handle the balance of the load.

- Consider computer programming of compression system.

- Consider use of VSD(s). Consult facility engineer when selecting appropriate VSD(s) and programming.

- In systems utilizing multiple reciprocating compressors, optimal performance can be realized by equal compressor unloading (to minimize suction line pressure drops to each compressor).

- When selecting screw compressor, consider one that will operate near its full capacity. Compare future replacement/adjustment costs to energy savings of getting properly sized compressor, when making a decision.

- Screw compressors should be used for base loading and reciprocating compressors should be used to meet the transient portion of a varying load [6].

- If two screw compressors are sharing a load, there is a point where it is better to fully load one compressor rather than split the load equally. In the case of two equally sized screw compressors, the optimal situation occurs when the compressors share the load up to an identifiable crossover point which occurs when the load on the system is about 65% of the combined available capacity of the compressors. Beyond that point it is best to fully load one of the screws and make up the difference with the other [6].

- When load sharing between two, unequal sized screw compressors is required, it is best to first fully load the smaller of the two, then at a certain identifiable crossover point, fully load the larger of the two compressors and make up the difference with the smaller of the two [6].

- Consider various software packages designed to minimize energy consumption.

- Single, larger compressor is typically less expensive. However, energy savings and reliability factor in favor of multiple compressor system.

- Parallel operation of two or more reciprocating compressors should be avoided unless there are strong and valid reasons for not using a single compressor. In a situation where two compressors must be used, extreme care in sizing and arranging the piping system is essential [6].

CONCLUSIONS

Compressors are complicated machines that require engineers to properly maintain them and determine savings opportunities. There are number of suggestions in this chapter which can help reduce the cost of using compressors in sustainable way.

Note: Related assessments found in appendix are: C, O, P, S, T, U, V.

REFERENCES

[1]   Natural Resources of Canada website. Available from: http://www.retscreen.net/ang/home.php. [Accessed March 2012].

[2]   Energy Savings with VSD Controlled Compressors. Available from: http://www.abb.com/cawp/seitp202/d8c203f60e3796adc12578a200580d94.aspx. [Accessed March 2012].

[3]   Choosing and Air Compressor. Available from: http://www.petersonpower.com/products/air-compressors-choosing.php. [Accessed March 2012].

[4]   Power Characteristics of Industrial Air Compressors. Available from: www.academic.udayton.edu/kissock/http/IAC/CompAir_ReducePres.doc. [Accessed March 2012].

[5]   What is the Optimum Compressor Discharge Pressure Set Point for Condensers? Available from: http://www.excaliburlpa.co.uk/downloads/Head%20Pressure%20NZ.pdf. [Accessed March 2012].

[6]   Load Sharing Strategies in Multiple Compressor Refrigeration Systems. Available from: www.irc.wisc.edu/file.php?id=50. [Accessed March 2012].

[7]   Calculating the True Savings when Reducing Air System Pressure. Available from: http://www.plantengineering.com/search/search-single-display/calculating-the-true-savings-when-reducing-air-system-pressure/c220531fc1.html. [Accessed March 2012].

[8]   Compressor and Compressed Air Systems. Available from: http://www.energyefficiencyasia.org/docs/ee_modules/Compressors%20and%20Compressed%20Air%20Systems.pdf. [Accessed March 2012].

[9]   Almasi A. Practical notes and latest technologies on modern centrifugal compressor component and system selection. Australian J. of Mech. Eng. 2012; 10-1: 71-80.

Send Orders of Reprints at reprints@benthamscience.net
Energy Assessments for Industrial Complexes, 2013, 95-97

CHAPTER 17

Belt Conveyors

Alexander Spivak, Ashok Kumar[*] and Matthew Franchetti

University of Toledo, Toledo, OH, USA

Abstract: Belt conveyors are generally optimized by engineers. However, there are number of ways that may reduce cost of running belt conveyors, predominantly through maintenance practices.

Keywords: Belt conveyors, belt conveyor selection, belt conveyor maintenance, aluminum frame conveyors, belt conveyors, cleated belt conveyors, modular conveyors for curves & turns, food quality conveyors, industrial conveyors, lift gate conveyors, low profile conveyors, magnetic belt conveyors, plastic chain conveyors, portable conveyors, stainless steel frame conveyors, steel frame conveyors, vacuum belt conveyors, z-frame conveyors, belt conveyor speed, belt conveyor friction.

INTRODUCTION

There are two main industrial classes of belt conveyors: material handling and bulk material handling. A belt conveyor consists of two or more pulleys (drives and idler) connected by a continuous loop belt. Powered (drive) pulley(s) moving the belt forward. A great number of specific types of conveyors are available, which include: aluminum frame conveyors, belt conveyors, cleated belt conveyors, modular conveyors for curves & turns, food quality conveyors, industrial conveyors, lift gate conveyors, low profile conveyors, magnetic belt conveyors, plastic chain conveyors, portable conveyors, stainless steel frame conveyors, steel frame conveyors, vacuum belt conveyors, z-frame conveyors, *etc.*

*Address correspondence to Ashok Kumar: Department of Civil Engineering, The University of Toledo; 3006 Nitschke Hall, Mail Stop 307, 2801 W. Bancroft St., Toledo, OH 43606, USA; Tel: 419-530-8136, Fax: 419-530-8116; E-mail: ashok.kumar@utoledo.edu

BELT CONVEYOR'S IMPROVEMENT OPPORTUNITIES

Generally, the improvement of energy efficiency of a belt conveyor system can be achieved through any one of its four components (performance, operation, equipment, and technology) [6]. There are number of ways to improve the sustainability of belt conveyors and improve the bottom line. They include [1-5]:

- Chose belt speed to be slightly higher than the belt speed required for nominal belt fill in order to reduce the risk of overloading in the case of small material flow fluctuations.

- Consider a high-efficiency gear reducer. Selection of helical bevel reducer may significantly increase the efficiency.

- Consider gear belts rather than roller chain drives.

- Minimize friction between the slide bed and the underside of the belt surface. Here are some expert advise [1-3]:

  o The lowest friction loading is obtained with a roller deck, which can be up to 100 times lower than a slide-bed conveyor.

  o For vibratory conveyors, the lowest energy use occurs when the conveyor is tuned and run at the spring system's natural frequency.

  o For auger conveyors, a high-efficiency gear reducer and a low-coefficient-of-friction plasma coating are two things that can reduce energy use.

- Optimize tension of mechanical assemblies.

- Consider operating mechanical conveyors just above the required production rate.

- Consider a variable-speed drive. Studies [5, 6] showed that VFD is in fact effective energy saving tool. However, use of VSD in constant speed motors is debatable. On-site engineers should make such decisions.

- Good control system can help optimization of conveyor's performance.

- When possible, fill all of the available conveying space. However, some products do not allow maximum fill. Consult on-site engineer for more details.

- Optimize the purge or clean-out cycles. Since purge cycles consume energy without transferring material, they're wasteful.

- Consider replacing smooth belt with notched belt where applicable.

CONCLUSIONS

Properly maintained belt conveyors are a good source of cost avoidance. Suggestions above may help manage and improve belt conveyors.

Note: Related assessments found in appendix are: C.

REFERENCES

[1]   Conveyor types and conveyor systems. Available from: www.dornerconveyors.com/conveyors. [Accessed March 2012].
[2]   Speed Control on Belt Conveyors: Does It Really Save Energy? Available from: www.synergy-eng.com/pdf/BSH-2005_Beltspeed_Lauhoff.pdf. [Accessed March 2012].
[3]   Reducing Energy Consumption on Overland Conveyors. Available from: http://www.womp-int.com/story/2009vol05/story028.htm. [Accessed March 2012].
[4]   Hiltermann J, Lodewijks G, Schott DL, Rijsenbrij JC, Dekkers JAJM, Pang Y. A Methodology to Predict Power Savings of Troughed Belt Conveyor by Speed Control. Particulate Sci. and Tech. 2011; 29: 14-27.
[5]   Zhang S, Xia X. Modeling and energy efficiency optimization of belt conveyors. Applied Energy 2011; 88: 3061-3071.
[6]   Zhang S, Xia X. Optimal control of operation efficiency of belt conveyor systems. Applied Energy 2010; 87: 1929-1937.

Send Orders of Reprints at reprints@benthamscience.net

CHAPTER 18

Water Heaters

Alexander Spivak, Ashok Kumar[*] and Matthew Franchetti

University of Toledo, Toledo, OH, USA

Abstract: Proper selection and good maintenance of a water heater may result in significant savings.

Keywords: Water heater, water heater maintenance, water heater selection, water heater efficiency, gas water heater, electric water heater, solar water heater, hybrid water heater, gas condenser, tankless water heater, point-of-use water heater, water heater insulation, type of water heater, water heater replacement, water heater insulation, flushing water heater, water heater baffle, cost of water heater, water heater energy consumption, when to replace water heater.

INTRODUCTION

There are a number of different types of water heating systems available on market. In order to optimize the costs of hot water, two main approaches may be considered: replacement of the water heater and good maintenance.

REPLACING WATER HEATER

A water heater is typically replaced only as needed. Generally, the savings generated by the replacement of a properly working unit is insufficient to justify replacement. However, once unit starts malfunctioning, replacement becomes a viable option. Efficiency improvement relative to the baseline design reduces the life cycle cost in the majority of homes for both gas and electric storage water heaters, and heat pump electric water heaters and condensing gas water heaters provide a lower life cycle cost for homes with large rated volume water heaters [3]. Similarly, businesses would benefit as much. There are number of different

*Address correspondence to Ashok Kumar: Department of Civil Engineering, The University of Toledo; 3006 Nitschke Hall, Mail Stop 307, 2801 W. Bancroft St., Toledo, OH 43606, USA; Tel: 419-530-8136, Fax: 419-530-8116; E-mail: ashok.kumar@utoledo.edu

water heater types that may be considered. Even though, it is generally less expensive to replace old water heater with a new one of the same type, other alternatives may have better payback period. The payback period is the difference in total investment cost divided by the difference in energy and maintenance savings between the two considered units. Because of the replacement costs, longevity should be included as part of the calculations. For example, one solar water heater will last as long as two gas storage tank units. When making replacement decisions, it is best to obtain information specific to units considered. Table **1** below presents various generic water heater options.

Table 1: Water Heater Types [1, 2]

Type	Solar Water Heater	Storage Tank (Gas)	Storage Tank (Electric)	Hybrid Electric Addition
Throughput	Varies	High	High	Medium - High
Efficiency	Very High (almost no waste)	Low (around 60%)	Low (around 60%)	High (around 85%)
Tax Credit	Yes	on Energy Star	on Energy Star	Yes
Initial Cost	High	Low - Medium	Low	High
Long Term Cost	Low	High	Very High	Low
Longevity	25 + yrs.	6 - 15 yrs.	6 - 12 yrs.	Varies
Maintenance	High	Low	Low	Medium - High
Reliability	Medium	High	High	Medium - High
Availability	Low, May need another small water heater to compensate for a string of cloudy days	High	High	Medium - High

Type	Gas Condenser Addition	Tankless	Point-of-Use
Throughput	High	Low - High	Low (good for small office)
Efficiency	Medium	High (around 80%)	High (around 80%)
Tax Credit	Yes	on Energy Star	on Energy Star
Initial Cost	Medium	High	High (per entire facility)
Long Term Cost	Low - Medium	Low - Medium	Low - Medium
Longevity	Varies	15 - 20 yrs.	Varies

Table 1: contd....

Maintenance	Medium - High	Medium - High	Medium - High
Reliability	Medium - High	Medium - High	Medium - High
Availability	Medium - High	Medium - High	Medium - High

There are other water heating options being developed. An example of newer water heating methods includes injected hydrogen combustion exhaust [5]. These and other novelties have not yet been proven in practice to be better than water heating systems discussed above, but may soon be considered as options.

MAINTENANCE OPTIONS

There are a few suggestions that may reduce the cost of ownership of water heaters [1, 2]:

- Consider direct-fired condensing hot water heaters for sanitizing hot water needs at food processing plants. They are 95-99% efficient.

- The water heater intake should have sufficient quantity and dispersion of the water.

- Flush tanks every 3 to 6 months to minimize sediment built-up.

- Consider a water heater blanket to insulate the tank.

- Insulate hot water pipes.

- Consider low flow shower heads.

- Consider use of baffle. It was shown to reduce natural gas consumption in residential water heater by 5% [4].

CONCLUSIONS

When making replacement decisions, it is worth considering alternative types of water heaters, as they may prove to have lower total cost of ownership. With proper maintenance water heaters are likely to serve significantly longer time.

Note: Related assessments found in appendix are: D, E, V.

REFERENCES

[1]   A Technical Guide to Designing Energy-Efficient Commercial Water Heater Systems. Available from: www.statewaterheaters.com. [Accessed March 2012].

[2]   High Efficiency Water Heaters. Available from: www.energystar.gov. [Accessed March 2012].

[3]   Levkov A, Franco V, Meyers S, Thompson L, Letschert V. Energy Efficiency Design Options for Residential Water Heaters: Economic Impacts on Consumers. ASJRAE Transactions 2011; 117-1: 103-110.

[4]   Moeini Sedeh M, Khodadadi JM. Energy efficiency improvement and fuel savings in water heaters using baffles. Applied Energy 2013; 102: 520-533 (in print).

[5]   Toja-Silva F. A novel water heater using injected hydrogen combustion exhaust. Energy and Buildings 2011; 43: 2320-2328.

Send Orders of Reprints at reprints@benthamscience.net

CHAPTER 19

Furnaces, Air Conditioners and Dehumidifiers

Alexander Spivak, Ashok Kumar[*] and Matthew Franchetti

University of Toledo, Toledo, OH, USA

Abstract: Furnaces, air conditioners and dehumidifiers are heavy users of energy. By properly insulating pipes, installing properly sized units, and considering various types of HVAC and dehumidifying devices, it is possible to save significant amounts of money while practicing sustainable business.

Keywords: Furnace, air conditioner, A/C, dehumidifier, HVAC, cost of heating, cost of cooling, heat cost reduction, heating systems, HVAC system optimization, HVAC decisions, furnace sizing, air conditioner sizing, heating, ventilation and air conditioning, AFUE, retrofitting of existing heating system, retrofitting boiler, air conditioning, oil furnace, gas furnace, electric furnace, propane furnace, wood burning furnace, corn burning furnace.

INTRODUCTION

This chapter discusses various heating, cooling and dehumidifying options, and advantages and disadvantages of each which will help business managers decide when to replace existing systems and what to replace them with.

HEATING SYSTEMS

General Information

Furnaces heat air and distribute the heated air using ducts. Boilers heat water, providing either hot water to baseboard radiators, radiant floor systems, coils, or steam to radiators. Steam boilers are generally less efficient than hot water boilers because they operate at higher temperatures. A central furnace or boiler's efficiency is measured by annual fuel utilization efficiency (AFUE) in residential houses or by combustion efficiency in commercial buildings.

*Address correspondence to Ashok Kumar: Department of Civil Engineering, The University of Toledo; 3006 Nitschke Hall, Mail Stop 307, 2801 W. Bancroft St., Toledo, OH 43606, USA; Tel: 419-530-8136, Fax: 419-530-8116; E-mail: ashok.kumar@utoledo.edu

AFUE is the ratio of heat output of the furnace or boiler compared to the total energy consumed by a furnace or boiler. AFUE doesn't include the heat losses of the duct system or piping, which can be as much as 35% of the energy for output of the furnace. When lower end furnaces and boilers are considered, the purchaser has to be wary of "fleet efficiency" ratings, which is the average efficiency of all models sold under a particular brand's name.

Since, an all-electric furnace or boiler has no flue loss through a chimney, the AFUE rating is between 95% and 100%, where losses are due to jacket heat loss of the units installed outside of the building. Below is a list of estimated AFUE values of the heating systems [1]:

Old, low-efficiency heating systems:

- Natural draft that creates a flow of combustion gases.

- Continuous pilot light.

- Heavy heat exchanger.

- 68%–72% AFUE.

- Older systems can drop into 50% range.

Mid-efficiency heating systems:

- Exhaust fan controls the flow of combustion air and combustion gases more precisely.

- Electronic ignition (no pilot light).

- Compact size and lighter weight to reduce cycling losses.

- Small-diameter flue pipe.

- 80%–83% AFUE.

High-efficiency heating systems:

- Condensing flue gases in a second heat exchanger for extra efficiency.

- Sealed combustion.

- 90%–97% AFUE.

Heating System Decisions

There are number of possible ways to save heating. Since annual cost of heat is typically around 30% of the cost of the heating system, there are a number of opportunities to save money. Such financial/sustainability solutions may involve retrofitting or replacing existing furnace or boiler. When unit is considered for replacement, proper sizing is very important. Too small unit will not be able to keep up with demand. The result may be shorter life span, uncomfortable temperatures and greater energy use. Too large system costs more to purchase, will cycle more frequently and result in higher energy cost and wear. In either case, improperly sized unit will result in economic loss.

AVAILABLE TECHNOLOGIES

Table **1** describes available technologies and their efficiencies.

Table 1: Energy by Source Type [1-4]

Energy Source (Energy Content)	Technology	Seasonal Efficiency (AFUE) %	Energy Savings % of Base*	Advantages	Disadvantages
Oil 138,700 BTU/gal	Cast-iron head burner (old furnace)	60	Base		1) Inefficient and expensive to operate 2) High carbon emissions
	Flame-retention head replacement burner	70–78	14–23		1) Higher carbon emission than gas
	High-static replacement burner	74–82	19–27	1) Efficient for low heat demand cases	1) Higher carbon emission than gas

Table 1: contd….

	New standard model	78–86	23–30	1) Can be as efficient as some gas furnaces	1) Higher carbon emission than gas
	Mid-efficiency furnace	83–89	28–33	1) Can be as efficient as some gas furnaces	1) Higher carbon emission than gas
	Integrated space/tap water (mid-efficiency)	83–89	28–33 space	1) May be economical to operate	1) Higher carbon emission than gas
			40–44 water	1) May be economical to operate	1) Higher carbon emission than gas
	1970-1992 efficiency estimate	65	7		1) Higher carbon emission than gas
Natural Gas 1,007 BTU/ft^3	Conventional	60	Base		1) Inefficient and expensive to operate
	Vent damper with non-continuous pilot light	62–67	3–10	1) Inexpensive	
	Mid-efficiency	78–84	23–28	1) Relatively inexpensive 2) Cost-effective in milder climates	1) Higher long term cost
	High-efficiency condensing furnace	89–97	33–38	1) May qualify for tax credit 2) Low long term energy costs 3) Low CO_2	1) High initial costs 2) Less cost effective in mild climate
	Multi-stage modulating gas furnace	92-97	34-38	1) Maintains constant temperature 2) Relatively cheap to operate	1) Expensive 2) Require new venting and other modifications
	Integrated space/tap water (condensing)	89–96	33–38 space		
			44–48 water		
	1970-1992 efficiency estimate	65%	7		

Table 1: contd….

Hybrid Gas Furnace / Electric Heat Pump				1) Very economical 2) Environmentally friendly 3) Available tax credit	1) High initial costs
Electricity 3,413 BTU/kWh	Electric baseboards (including older units)	100			
	Electric furnace or boiler	100			
	Air-source heat pump	1.7 COP**			
	Earth-energy system	2.6 COP**		1) Environmentally friendly 2) Energy efficient	1) May not be quiet 2) Initially expensive
	(ground-source heat pump)				
Propane 92,700 BTU/gal	Conventional (including older units)	62	Base	1) Propane is more efficient than gas	
	Vent damper with non-continuous pilot light	64–69	3–10	1) Propane is more efficient than gas	
	Mid-efficiency	79–85	21–27	1) Propane is more efficient than gas	
	Condensing	87–94	29–34	1) Propane is more efficient than gas	
Wood Hardwood - 28,000,000 BTU/cord Softwood - 17,000,000 BTU/cord Wood pellets - 20,000,000 BTU/cord	Central furnace	45–55		1) Uses renewable source	
	Conventional stove (properly located)	55–70		1) Uses renewable source	
	"High-tech "stove*** (properly located)	70–80		1) Very inexpensive to operate 2) Uses renewable source	
	Advanced combustion fireplace	50–70		1) Uses renewable source	
	Pellet stove	55–80		1) Uses renewable source	

Table 1: contd….

	Non-airtight stove	20-30		1) Uses renewable source	
	Fireplace	5		1) Uses renewable source	
Corn Shelled corn - 7,000 BTU/lb	Stove (some can be modified as a fireplace insert)	55 - 85		1) Very inexpensive to operate 2) Uses renewable source	
Straw - 6,550 BTU/lb Corn Stover -7,540 BTU/lb	Hot air furnace	55 - 85		1) Very inexpensive to operate 2) Uses renewable source	

Important note: higher efficiency systems are best justified in colder climates only.
* "Base" represents the energy consumed by a standard furnace.
** COP =Coefficient of performance, a measure of the heat delivered by a heat pump over the heating season per unit of electricity consumed.
*** CSA B415 or EPA Phase II tested.

Presented efficiencies in the table are established over the years. However, new energy efficiency indicator methods [9] as well as new methods for determination of efficiency are being continuously developed.

One technology not described above is the Micro-cogeneration system. This is a newer, not yet proven technology, which may provides high efficiency heating, partial grid independence or re-powering grid and it may store heat for later need. However, it is new to market therefore may be very expensive and unreliable. This system may prove to be very advantageous, but must be approached with caution.

In order to determine which technology is best for the specific facility, compare current system with other (better) systems within the same category. Using "Energy savings %" column, estimate percent difference between current and better systems. Then, multiply current annual heating cost by percent difference. This value is estimated annual savings generated by replacing current system with a new system. Calculate the simple payback period by dividing total cost of the new system by annual savings. Table **2** below can help determine the cost savings. The use of this table, however, is limited and specific to condition of the property, length and condition of piping system, and the specifics of the heating system itself.

Table 2: Efficiency and Cost of Use of Heating Equipment [1]

Annual Estimated Savings for Every $100 of Fuel Costs by Increasing Heating Equipment Efficiency (assuming the same heat output)									
Existing System AFUE	New / Upgraded System AFUE								
	55%	60%	65%	70%	75%	80%	85%	90%	95%
50%	$9.09	$16.76	$23.07	$28.57	$33.33	$37.50	$41.24	$44.24	$47.36
55%		$8.33	$15.38	$21.42	$26.66	$31.20	$35.29	$38.88	$42.10
60%			$7.69	$14.28	$20.00	$25.00	$29.41	$33.33	$37.80
65%				$7.14	$13.33	$18.75	$23.52	$27.77	$31.57
70%					$6.66	$12.50	$17.64	$22.22	$26.32
75%						$6.50	$11.76	$16.66	$21.10
80%							$5.88	$11.11	$15.80
85%								$5.55	$10.50
90%									$5.30

When considering replacing a heating source, take into account as many systems as is practical. Best purchasing and operational costs estimates may be obtained from manufacturers.

PERFORMANCE OPTIMIZATION

Performing regular maintenance on existing systems can also optimize performance and reduce costs. Periodically check the following:

All systems [1, 5, 6]:

- Check condition of the chimney and connection with a pipe. Newer systems with 90%+ efficiency do not require chimney.

- Check the physical integrity of the heat exchanger. Leaky boiler heat exchangers leak water and are easy to spot. When furnace heat exchangers leak, they mix combustion gases with the air we breathe— an important safety reason to have them inspected.

- Adjust the controls on the boiler or furnace to provide optimum water and air temperature settings for both efficiency and comfort. A programmable thermostat will typically pay for itself very quickly.

- If you're considering replacing or retrofitting your existing heating system, consider a combustion-efficiency test.

Forced-air Systems [1, 7]:

- Check the combustion chamber for cracks.

- Test for carbon monoxide (CO).

- Adjust blower control and supply-air temperature (typically manual adjustments are performed).

- Clean and oil the blower.

- Remove dirt, soot, or corrosion from the furnace or boiler. This task may be done with mini vacuum cleaner.

- Check fuel input and flame characteristics, and adjust if necessary.

- Seal connections between the furnace and main ducts. High temperature caulk and tape are available for this task.

Hot-water Systems [1]:

- Test pressure-relief valve.

- Test high-limit control.

- Inspect pressure tank, which should be filled with air, to verify that it's not filled with water.

- Clean the heat exchanger.

Steam Systems [1]:

- Drain water from the boiler to remove sediments. This improves the heat exchange efficiency.

- Test low-water cutoff safety control and high-limit safety control.

- Drain the float chamber to remove sediments. This prevents the low-water cutoff control from sediment clogs.

- Analyze boiler water and add chemicals as needed to control deposits and corrosion.

- Clean the heat exchanger.

Many older chimneys have deteriorated liners or no liners at all and must be relined during furnace or boiler replacement. A chimney should be relined when any of the following changes are made to the combustion heating system:

- When you replace an older furnace or boiler with a newer one that has an AFUE of 80% or more. These mid-efficiency appliances have a greater risk of depositing acidic condensation droplets in chimneys, and the chimneys must be prepared to handle this corrosive threat. The new chimney liner should be sized to accommodate both the new heating appliance and the combustion water heater by the installer.

- When you replace an older furnace or boiler with a new 90+ AFUE appliance or a heat pump. In this case, the heating appliance will no longer vent into the old chimney, and the combustion water heater will now vent through an oversized chimney. This oversized chimney can lead to condensation and inadequate draft. The new chimney liner should be sized for the water heater alone, or the water heater in some cases can be vented directly through the wall.

SIZING THE SYSTEM

Decisions on how much heating or cooling is needed is also a factor of the climate zone as well as condition of the property. Size of the heating unit may be calculated using the Table **3**.

Table 3: Heating and Cooling Requirements Per Climate Zone [1]

Climate Zone		Heating Requirement in BTU / Square Foot				
		Zone 1	Zone 2	Zone 3	Zone 4	Zone 5
Heating requirement of well insulated structure	---	30	35	40	45	50
Heating requirement of poorly insulated structure	---	35	40	45	50	60
Climate zone		Size of the Structure/Zone (in Square Feet)				
	A/C unit size	Zone 1	Zone 2	Zone 3	Zone 4	Zone 5
Cooling requirement of well insulated structure	1.5 Tons	600	600	600	700	700
Cooling requirement of poorly insulated structure	1.5 Tons	900	950	1000	1050	1100
Cooling requirement of well insulated structure	2 Tons	900	950	1000	1050	1100
Cooling requirement of poorly insulated structure	2 Tons	1200	1250	1300	1350	1400
Cooling requirement of well insulated structure	2.5 Tons	1200	1250	1300	1350	1400
Cooling requirement of poorly insulated structure	2.5 Tons	1500	1550	1600	1600	1650
Cooling requirement of well insulated structure	3 Tons	1500	1500	1600	1600	1650
Cooling requirement of poorly insulated structure	3 Tons	1800	1850	1900	2000	2100
Cooling requirement of well insulated structure	3.5 Tons	1800	1850	1900	2000	2100
Cooling requirement of poorly insulated structure	3.5 Tons	2100	2150	2200	2250	2300
Cooling requirement of well insulated structure	4 Tons	2100	2150	2200	2250	2300
Cooling requirement of poorly insulated structure	4 Tons	2400	2500	2600	2700	2700
Cooling requirement of well insulated structure	5 Tons	2400	2500	2600	2700	2700
Cooling requirement of poorly insulated structure	5 Tons	3000	3100	3200	3300	3300

Note: there are other factors such as facility type, style of the door, number of customers, internally produced heat, *etc.* that affect these values. These values are intended as a guideline only.

Proper sizing of the system will reduce energy consumption, improve comfort level, increase life of the system and reduce downtime and maintenance costs.

OTHER ECONOMICAL/SUSTAINABLE OPTIONS

There are number of simple no- or low-cost methods of reducing the utility bill. They include:

- Consider programmable thermostat with program that makes sense.

- Keeping temperature at 74°F in summer and 68°F in winter.

- Consider zoned heating, especially if some areas of the facility require less heat that other areas. Consider isolation and insulation of areas with various temperature demands.

- Give special attention to the economizer.

- Give special attention to ductwork.

ECONOMIZERS

Cooling system that have a capacity of 7.5 tons or more will typically include economizers. A properly operating economizer can cut energy costs by as much as 10% of a building's total energy consumption (up to 20% in mild, coastal climates), depending on local climate and internal cooling loads.

However, many newly installed economizers don't work properly and the majority of the older units actually malfunction. As result, they consume energy instead of saving it. If an economizer breaks down when its damper is in open or nearly open position, peak loads increase as cooling or heating systems try to compensate for the excess air entering the building. There are number of economizer-related recommendations:

- Test economizers from the start and at least twice a year thereafter. Here are three testing techniques;

- o   Observe the damper position – make sure it is consistent with setting.

- o   Fool the economizer controls – on the cool day warm the sensor with your hand and see if economizer damper moves to closed/near closed position.

- o   Install temperature data loggers.

- If repairs are impractical, lock the economizer in the minimum-outside-air position. This will allow for least amount of loss through economizer. Alternatively, if possible, remove the economizer.

- Some economizers cannot be cost-effectively maintained if the are located in corrosive environment, not able to produce substantial savings, or installed in buildings with undersized outside/exhaust air openings.

DUCTWORK

Proper ductwork can make a large difference in heating and cooling bills. Leaks in the ductwork may cause significant drops in efficiency of the HVAC system. Leaks may be physically felt by hand when system is operating. When searching for leaks, inspector should pay close attention to joints. Leaks may be observed by use of soap water, special bubbling solution or electronic leak detector.

Ductwork loses significant amounts of heat if not properly insulated. Selection of insulation materials depends on condition the ductwork is exposed to and type of the ductwork. The general recommendation is to insulate ductwork completely to the point where exterior temperature of the insulation is close to the outside air temperature.

DEHUMIDIFIERS

One final note for consideration is a desiccant-based dehumidification system to replace part of cold-coil dehumidification. Such replacements typically have payback periods less than eight years. Below are some advantages of the replacement system [8].

- Refrigeration-based dehumidification systems are more economical than desiccants at high temperatures and high moisture levels. In general, mechanical refrigeration systems are seldom used for applications below 45% RH.

- Desiccant-based systems are more economical than refrigeration systems at lower temperatures and lower moisture levels. Typically, a desiccant dehumidification system is utilized for applications below 45% RH down to 1% RH.

- The difference in the costs of electrical power and thermal energy (*i.e.,* natural gas or steam) will determine the ideal mix of desiccant to refrigeration-based dehumidification in a given application.

  o If thermal energy is cheap and power costs are high, a desiccant based system will be most economical to remove the bulk of the moisture from the air.

  o If power is inexpensive and thermal energy for reactivation is costly, a refrigeration based system is the most efficient choice.

- Certain desiccant dehumidifiers may benefit from use of Synchronous belt drives. However, synchronous belt drives are less durable than V-drives. In addition, retrofitting V-drive may require additional system strengthening [10].

CONCLUSIONS

When it comes to heating, cooling and dehumidifying there are many options to consider. If properly studied, it is possible to save a significant amount of money by selecting the best option and carrying out proper system maintenance.

Note: Related assessments found in appendix are: B, C, I, J, K, N, R, T, V.

REFERENCES

[1]   US Department of Energy website. Available from:
       http://www.energysavers.gov/your_home/space_heating_cooling/index.cfm/mytopic=1253
       0. [Accessed March 2012].

[2]   AC4Life website. Available from: http://www.acdirect.com/systemsize.php. [Accessed March 2012].

[3]   Furnace Oversizing Facts [Recurve website]. Available from: http://blog.recurve.com/furnace-oversizing-facts. [Accessed March 2012].

[4]   HVAC: Economizers [Reliant website]. Available from: http://www.reliant.com/en_US/Platts/PDF/P_PA_8.pdf. [Accessed March 2012].

[5]   A Home Inspectors' Guide to Flexible HVAC Duct Installation [ASHI Reporter website]. Available from http://www.ashireporter.org/articles/articles.aspx?id=41. [Accessed March 2012].

[6]   NetWorx website. Available from: http://hvac.networx.com/info/ductwork-insulation/. [Accessed March 2012].

[7]   NDSU website. Available from: http://www.ag.ndsu.edu/pubs/ageng/structu/ae1015a.pdf. [Accessed January 2012].

[8]   Desiccant Dehumidification *vs.* Mechanical Refrigeration [Bry-Air website]. Available from: http://www.bry-air.com/applications/moisture_regain_prevention/desiccant-dehumidification-*vs.*-mechanical-refrigeration.html. [Accessed March 2012].

[9]   Perez-Lombard_L, Ortiz J, Maestre IR, Coronel JF. Constructing HVAC energy efficient indicators. Energy and Buildings 2012; 47: 619-629.

[10]   Griffiths B. Optimizing the Energy Efficiency of Desiccant Dehumidifiers. Heating / Piping / Air Conditioning Engineering 2011; 83-12: 38-42.

CHAPTER 20

Heat Recovery

Alexander Spivak, Ashok Kumar[*] and Matthew Franchetti

University of Toledo, Toledo, OH, USA

Abstract: Some facilities may have heat recovery options available. These options may include parts washers, boilers, and any machines that produce significant recoverable heat. Often, such projects are expensive, but may still have reasonable payback periods.

Keywords: Heat recovery, heat redistribution, exothermic process, heat transfer, parts washer, office heating, heat redistribution project.

INTRODUCTION

Heat generated in one section of a facility may be harvested and used in another section. For example, unused heat from a parts washer may be harvested and used to heat nearby offices.

GENERAL CONCEPT

There are a number of the exothermic processes from which the energy may be harvested and used. The payback period of the heat recovery system is calculated by addition of energy savings up to the point when savings are equal to the system's cost. Areas of the plant that can be considered include but not limited to parts washers, heat exchangers, boilers, motors, and chemical reactors.

FEASIBILITY AND COST ANALYSIS OF HEAT RECOVERY SYSTEMS

The decision to recover heat is based on need of heat, heat availability and cost of heat recovery. The first step in making such decision is to determine how much heat the exothermic process generates. The next step is to determine savings

---

*Address correspondence to Ashok Kumar: Department of Civil Engineering, The University of Toledo; 3006 Nitschke Hall, Mail Stop 307, 2801 W. Bancroft St., Toledo, OH 43606, USA; Tel: 419-530-8136, Fax: 419-530-8116; E-mail: ashok.kumar@utoledo.edu

associated with use of this heat. Savings maximum is equal to cost of heating of the areas that may be heated from harvested energy. The savings maximum must be adjusted by answering following questions:

- Will recovered heat be supplemented by other heating source (for example, recovered heat is insufficient or exothermic process is not steady state)?

- Is there a less expensive way to reduce heating cost and possibly cooling cost by insulation improvement, window replacement, *etc.*?

- Will use of recovered heat effectively oversize current HVAC system?

If it is determined that heat transfer system is feasible, the next step is to decide whether it is cost effective. An engineer will have to optimize heat loss across the system and determine how much heat will be delivered, how effectively heat may be controlled and what is a payback period of such system.

Alternative type of heat recovery system is a supplementary system. Such system supplements existing heating system of the facility. These systems are usually more expensive to install, but may be more cost efficient in the long run and more convenient to use.

CONCLUSION

Heat recovery options feasibility studies are typically done by engineers. Any source of significant heat that is currently not being utilized may be considered.

Note: Related assessments found in appendix are: H, J.

Send Orders of Reprints at reprints@benthamscience.net

CHAPTER 21

Documents

Alexander Spivak, Ashok Kumar[*] and Matthew Franchetti

University of Toledo, Toledo, OH, USA

Abstract: Replacement of paper documents with computer files and reduction/elimination of paper-based correspondence are common sustainability methods that are usually cost-effective.

Keywords: Fiber recycling, office documents, paper recycling, economic printing, employee exchange, documents, electronic documents, electronic document substitution, electronic filing, office management, green office, paperless, paperless office, paper-based correspondence, paperless correspondence, electronic files, e-files, e-file management, document reduction, document rotation.

INTRODUCTION

Many companies are switching to paperless or reduced paper operations to save money and become more economically sustainable. In order to reduce paper use, companies may use fax-to-email to reduce or eliminate faxes, using e-mails and their website for communication to customers and vendors, introduce double-sided printing, circulate a single copy of a document for proofreading, if needed utilize two computer monitors per station to eliminate need to have paper versions of reviewed documents (there is an environmental and economic cost to this approach – a careful decision should be made prior to allowing two computer monitors), receive online versions of magazines and other literature, and use old, non-confidential documents as scratch paper. Often, companies involved in such activities also have websites to promote green activity, which include "green updates", an employee suggestion box or forum, and employee exchange page (where employees may trade personal items).

DOCUMENT REDUCTION AND PUBLIC RELATIONS

Going "green" is often an economic decision. However, it is not an uncommon

---

*Address correspondence to Ashok Kumar: Department of Civil Engineering, The University of Toledo; 3006 Nitschke Hall, Mail Stop 307, 2801 W. Bancroft St., Toledo, OH 43606, USA; Tel: 419-530-8136, Fax: 419-530-8116; E-mail: ashok.kumar@utoledo.edu

practice to use document reduction in public relations. Techniques commonly used by firms include but not limited to website announcements, press releases and participation in public events. The potential revenue due to attraction of new customers may be considered as part of document reduction decision. Many sustainable activities result in greater environmental benefits than document reduction, however, the latter is among the most "visible" activities.

DOCUMENT REDUCTION OPTIONS

One of the effective ways of saving money is to reduce the use of paper and other paper products. File backup systems are now more reliable than ever, reducing the need to retain paper copies of documents. Paper copies storage space and maintenance may cost hundreds of dollars per month.

In addition to reduced need for storage, company may introduce a policy, which asks employees to reduce amount of printed paper and print on both sides of the page, when feasible.

On the company's side, policies of replacing paper documents with scanners, tablets, or additional computer monitors may be considered, but may not be economical unless large numbers of shipping/receiving or inventory documents are used. In addition, a company may choose to subscribe to electronic versions of magazines and newspapers, as opposed to paper versions.

One drawback to paperless conversion is the need to maintain security of the documents. Additional costs for firewalls, virus protection, and so on may need to be considered as well. The simple payback calculations vary, but typically are not involved.

CONCLUSION

By going paperless, companies can save time and money, and improve their relationships with customers.

Note: Related assessments found in appendix are: F, W.

Send Orders of Reprints at reprints@benthamscience.net

CHAPTER 22

Solid Waste Management

Alexander Spivak, Ashok Kumar* and Matthew Franchetti

University of Toledo, Toledo, OH, USA

Abstract: Solid waste management cost reduction is generally achieved by removing recyclables from the solid waste stream. The savings are observed in reduced cost of solid waste hauling. The recyclable collection may turn profitable as well.

Keywords: Solid waste, solid waste management, recycling, recycling methods, aluminum cans, plastic, paper, metal, cost of recycling, solid waste reduction, paper, glass, cardboard, OCC, magazine, newspaper, newsprint, ONP, solid waste hauling, bailer, compactor.

INTRODUCTION

Solid waste recycling may be a sustainable and economic way of managing waste. Recycling aluminum (cans) will likely generate reasonable profit both from profits gained from recycling and money saved by reducing solid waste. In addition, plastics, glass, metals and fibers may be recycled as well. Generally, plastics, glass and metal recycling will at least pay for itself. Fiber recycling (paper, fiber-based cardboard, magazines, newspaper) typically will pay for itself if it is subsidized by the local government program. Many counties and cities have recycling programs, so contacting local government is a good idea. Recycling will reduce overall solid waste load and reduce a company's solid waste hauling costs.

OTHER WASTE REDUCTION OPTIONS

An additional option for hauling cost reduction may be a bailer or compactor. A bailer or compactor can reduce hauling costs by as much as 75% and pay for itself within 5 to 15 years, depending on the application.

*Address correspondence to Ashok Kumar: Department of Civil Engineering, The University of Toledo; 3006 Nitschke Hall, Mail Stop 307, 2801 W. Bancroft St., Toledo, OH 43606, USA; Tel: 419-530-8136, Fax: 419-530-8116; E-mail: ashok.kumar@utoledo.edu

Wood pallets at the end of their useful life may also be recycled. Some mulch producers will accept wood pallets and may even pay for shipping. Newer pallets are often circulated between companies and pallet dealers. It is typically less expensive for larger companies to own rather than rent pallets.

Used packaging material is often accepted by shipping companies free or nearly free of charge and often collected without charge. If a company is involved in regular shipping, it may be advantageous to store packaging materials and cardboard boxes in a designated area for collection.

A large number of other items that are typically disposed of may be reused instead. Thrift stores collect old furniture for resale, schools and other non-profit organizations collect old computers, and energy recovery facilities can collect other unusable goods. Through effort, many companies are able to achieve a zero landfill system.

CONCLUSIONS

Solid waste recycling is usually both sustainable and economically feasible. Often free recycling assessments are available from the solid waste management division of the local government.

Note: Related assessments found in appendix are: Q, W.

Send Orders of Reprints at reprints@benthamscience.net

CHAPTER 23

Application of Calculators for the Energy Assessment

Srikar Velagapudi, Ashok Kumar*, Alexander Spivak and Matthew Franchetti

University of Toledo, Toledo, OH, USA

Abstract: This chapter emphasizes the importance of applying the calculators for assessing the energy usage and pollution prevention for a hypothetical facility in Ohio. As a part of the study, we identified lighting fixtures and suggested alternatives for energy reduction and pollution prevention. We performed an energy assessment survey of the entire facility using the Energy Assessment Spreadsheet (EAS) tool developed at The University of Toledo to calculate reductions in energy use and costs associated with energy consumption of lighting fixtures. As a result, there is a reduction of electricity usage from 2,056,942 kilo Watt hour (kWh) to 1,477,059 kWh, or nearly 30% less energy expenses with an annual savings of over $55 thousand. Additionally, we performed a life cycle assessment using Economic Input Output Life Cycle Assessment (EIO-LCA) to determine the change in environmental impacts due to the use of energy efficient lighting. EIO-LCA revealed a reduction in Carbon Dioxide (CO_2) gas emissions from 1,830 metric tons to 1,300 metric tons, which is approximately 30%. The pollution prevention (P2) calculator comprising of P2 Green House Gas (GHG) tool and the P2 Cost Calculator developed by the United States (U.S.) Environmental Protection Agency (EPA) was the third tool applied. The greenhouse gas reductions were calculated using the P2 GHG tool while cost savings related to energy savings were calculated using P2 cost calculator. There was a reduction of 492.823 metric tons of Carbon Dioxide equivalents (MTCO2e) GHGs for the electricity savings of nearly 580,000 kWh.

Keywords: Energy assessment, reductions in energy, life cycle assessment, energy consumption, environmental impacts, energy efficient lighting, pollution prevention calculator, P2, greenhouse gas reductions, electricity savings, reduction in hazardous waste generation, lighting fixtures, MSDS, LEAN, HVAC, HVAC Design: SBT: HAT, DGP, SBSAT, D-HAT, CIS, FAT, UT PPIS.

INTRODUCTION

Pollution prevention, and its affiliation with sustainability, has gained importance

*Address correspondence to Ashok Kumar:** Department of Civil Engineering, The University of Toledo; 3006 Nitschke Hall, Mail Stop 307, 2801 W. Bancroft St., Toledo, OH 43606, USA; Tel: 419-530-8136, Fax: 419-530-8116; E-mail: ashok.kumar@utoledo.edu

with the increase in regulatory policies enforced by various environmental agencies throughout the globe. The U.S. EPA defines P2 as the exercise of practices that reduce or eliminate the creation of pollutants by increased efficiency in the use of raw materials, energy, water, and other resources; and protection of natural resources by conservation [1]. The U.S. Pollution Prevention Act of 1990 states that pollution should be prevented or reduced at the source whenever feasible. The increase in consumer demand, along with environmental challenges, has made P2 an important factor. According to economic outlook 2010, activities in buildings such as heating, cooling, lighting, *etc.* accounts for about one fifth of the total delivered energy in the world [2]. It is predicted that the world's residential energy usage will increase by 1.1% per year from 2008 to 2035. Much of this growth pertains to the improvement of living standards [3]. Several countries have adopted procedures in achieving 'low energy' and 'zero energy' buildings. Energy certificates or building energy ratings for buildings can provide the potential energy performance of a particular building [4, 5]. A conventional building can vary its total amount of energies, direct and indirect required to build and operate over its life cycle (50 year-period) from 2% to 38% [6]. The total amount of energies, direct and indirect, for the entire amount of production

A building's Energy performance can be calculated using tools such as Energy Plus, developed by the Department of Energy (DOE), which gives detailed energy use on a sub-hourly basis [7]. Several environmental agencies and educational institutions have developed various P2 tools. The Air Pollution Research Group (APRG) of the Civil Engineering Department at The University of Toledo (UT)developed several user-friendly P2 tools, such as material safety data sheets v1.1/1.2 (MSDS) [8]; the lean assessment screening tool v1.0 (LEAN) [9]; the hybrid heating, ventilation, and cooling (HVAC) system design tool v1.0 (HVACDesign) [10]; the building sustainability tool v1.0 (SBT) [11]; the hospital assessment tool v1.0 (HAT) [12]; the database for green products v1.0 (DGP) [13]; the small business self-assessment tool v1.0 (SBSAT) [14]; the department specific hospital assessment tool v1.0 (D-HAT) [15]; chemical identification software v1.0 (CIS) [16]; and the food assessment tool v1.0 (FAT) [17]. These tools assist environmental managers to determine the opportunity for P2 savings and assess compliance with regulations. They can be accessed from the UT PPIS website (Available http://www.eng.utoledo.edu/aprg/ppis/ppistools.htm [18]). P2 deals with industry related issues such as modifications in process, recycling within the processes, managerial practices, and modifications to equipment. Kumar *et al.* [19] extensively studied the various aspects of P2 activities.

This chapter focuses on the assessment of energy efficiency, energy savings, and pollution prevention for a hypothetical facility in Ohio using P2 tools. Additionally, this chapter shows how one can use the EIO-LCA tool to determine the change in environmental impact after implementation of the recommendations in association with the use of the P2 tools. The tools used are: EAS v1.0 developed by the APRG at UT, with funding from the U.S. EPA, Available the UT PPIS Website (Available http://www.eng.utoledo.edu/aprg/ppis/ppistools.htm [18]); P2 Measurement Tools, developed by the U.S. EPA, comprised of two spreadsheets: the P2 Cost Calculator and the P2 GHG Calculator Available the U.S. EPA Website (Available http://www.epa.gov/p2/pubs/resources/measurement.html [20]); and the EIO-LCA tool, which was an internet-based tool developed by the Carnegie Mellon University Green Design Institute through partial funding from the National Science Foundation (NSF), the U.S. EPA, and the Green Design Consortium is available online for use by environmental professionals at EIOLCA Website (Available http://www.eiolca.net/index.html [21]).

Energy Assessment Spreadsheet v1.0 (EAS)

The Energy Assessment Spreadsheet tool is a simple and excellent platform for compiling input data and carrying out needed mathematical operations, which is an imminent part of the energy assessment project. The EAS has three sections, lighting, motors, and HVAC Systems, that account for the majority of industrial energy consumption. The lighting section is further divided into three sub-sections: input data, lighting cost, and lighting cost reductions. Inputting the wattage, quantity, annual operating hours for each category of fixtures, and the unit cost of energy automatically generates information about the annual energy consumption. It also reflects the operational cost of all individual fixture types and their corresponding totals. It is a very user friendly tool that provides an opportunity for identifying energy savings.

P2 Measurement Tools

The US EPA designed the P2 tools in a spreadsheet format for measuring the P2 activities from an environmental and economic perspective. The two spreadsheets, P2 GHG Calculator and P2 Cost Calculator, which are a part of the measurement tools under discussion, demonstrate a unique multi-media perspective in reducing GHG emissions and produce cost savings. The P2 GHG Calculator computes GHG emission reductions from electricity conservation, green energy, fuel and chemical substitutions with lower GHG-intensities, water conservation, and improved materials and process management. The P2 GHG Calculator enables the user to submit high quality data to the program and can be used in reducing MTCO2e. The P2 Cost Calculator determines the total cost savings associated with the conservation efforts.

Economic Input Output Life Cycle Assessment Tool (EIO-LCA)

EIO-LCA is an internet based tool that estimates the materials and energy resources required for, and the environmental emissions resulting from, activities in an economy. This tool evaluates the environmental impacts of a product or a process over its entire life cycle, and estimates the total emissions during a life cycle based on the information from industrial transactions, such as the purchase of materials between two or more industries and information regarding the environmental emissions from an industry. It analyzes the different sectors of an economy and takes into account the influences of indirect suppliers, unlike the traditional life cycle assessment procedures. While performing a LCA, the user needs to choose an appropriate model, industry and sector, monetary amount for the chosen sector, and category of results to obtain as an output from the tool.

The BC Corporation - R&D Headquarters

The appendix includes multiple examples of assessments and their results similar to the one shown below for the BC Corporation. The example below demonstrates application and effects of the energy assessment as well as sustainable energy policy improvement plan. As a part of this hypothetical energy assessment, we selected The BC Corporation's Research & Development headquarters. BC Corporation employs about 70,000 employees worldwide and sells products to customers in more than 150 countries. It is a global technology leader in electrical components and systems for power quality, distribution, and control. The entire R&D facility is divided into 4 areas: office, conference rooms, manufacturing, and miscellaneous. Entryways, storage areas and restrooms were considered miscellaneous.

Table 1: Energy Star Recommended Lighting Fixtures [22]

Incandescent Bulbs (watts)	Minimum Light Output (lumens)	Common Energy Star Qualified Bulbs (watts)
40	450	9 to 13
60	800	13 to 15
75	1100	18 to 25
100	1600	23 to 30
150	2600	30 to 52

Table 2: Comparison of Energy Star Recommended Fixtures to the Regular Fixtures [23]

Application and Example		Efficiency in Application per Technology		
		Incandescent	Halogen	HID
Low ceiling	Offices, conference rooms, classrooms	Low	Low	Medium
High ceiling	Warehouses, lobbies	Low	Low	Medium-High
Exterior lighting	Parking lots	Low	Low	Medium-High
Accent lighting	Displays, artwork	Low	Low-Medium	Medium
Task lighting	Desk lamps	Low	Low-Medium	Medium
Exit sign lighting	Exit signs	Low	N/A	N/A

Application and Example		Efficiency in Application per Technology		
		T12 Fluorescent (linear)	T8 Fluorescent (linear)	T5 Fluorescent (linear)
Low ceiling	Offices, conference rooms, classrooms	Medium	High	High
High ceiling	Warehouses, lobbies	Medium	Medium-High	High
Exterior lighting	Parking lots	Medium	High	High
Accent lighting	Displays, artwork	Medium	High	High
Task lighting	Desk lamps	Medium	High	High
Exit sign lighting	Exit signs	N/A	N/A	N/A

Table 2: contd….

Application and Example		Efficiency in Application per Technology	
		Compact Fluorescent	LED
Low ceiling	Offices, conference rooms, classrooms	Medium	N/A
High ceiling	Warehouses, lobbies	Medium-High	N/A
Exterior lighting	Parking lots	Medium	N/A
Accent lighting	Displays, artwork	High	High
Task lighting	Desk lamps	High	High
Exit sign lighting	Exit signs	Medium	High

EXPERIMENTAL METHODS

The energy assessment consisted of a comprehensive examination of current lighting fixtures and appliances in the entryways, conference rooms, and the office, manufacturing, and kitchenette areas. We carried out a lighting survey at the facility, involving study of lighting fixtures along with measurement of effectiveness of fixtures. We recorded information about each lighting fixture throughout the facility for the type, number, wattage, and the total number of hours each fixture operated in a year. The EAS then provided an estimate on the usage cost per year at $0.095/kWh for existing fixtures. Table **1** provided identification for alternative Energy Star lighting fixtures with the same amount of light output. For a particular wattage of an incandescent bulb the corresponding energy star qualified bulb was selected, while in the case of a fluorescent bulb the available minimum wattage energy star qualified bulb was selected. Observe from Table **2** that a T5 fluorescent bulb is a high efficiency bulb which can be installed at any location in a building except for the exit signs. The cost estimation was carried out for the replaced energy star fixtures using EAS. The P2 GHG calculator and the P2 Cost Calculator helped to determine the GHG reduction and dollar savings based on the total kWh of electricity conserved for the facility, respectively. The LCA was carried out using the EIO-LCA tool for the cases of before and after implementation of the recommendations. The "power generation and supply" sector was selected in the operational phase to run the EIO-LCA tool.

ENERGY ASSESSMENT FINDINGS

The BC Corporate facility has approximately 1,220 lighting fixtures With 631 in the office, 49 in the conference rooms, 478 in manufacturing, and 62 in miscellaneous areas. Each type of bulb was assigned an alphabetical designation for calculation purposes (Table **3**). The letters A, B and C correspond to office areas, D, E and F to conference rooms, G, H and I to manufacturing areas, and J, K and L to miscellaneous areas. To provide a clear view of how much energy efficiency can be attained, the data was analyzed based on type of area and total estimated savings. The facility had mostly T-8 and T-12 lighting fixtures which were replaced with recommended Energy Star lighting fixtures in accordance with Table **1**. For example, the fixture type 'E', a 135 watt incandescent as shown in the Table **3**, was replaced with a 30 watt fluorescent bulb, as shown in Table **5**. Similarly, fluorescent fixtures were replaced with respective energy star qualified fixtures.

Table **3** shows the number of existing lighting fixtures that were present in each area of the facility and their operating hours per year. Table **4** gives the annual costs calculated for the existing lighting fixtures at over $195,000 for the electricity usage of over 2 million kWh. Table **5** shows the number of replacement Energy Star lighting fixtures in each area of the facility and their operating hours per year. Table **6** gives the annual costs calculated for the Energy Star lighting fixtures, with the annual cost for the recommended lighting fixtures totaling just over $140,000 for the electricity usage of less than 1.5 million kWh.

The total cost reductions were calculated using the percentage reduction formula shown below. Table **7** tabulates the total savings per year over $55,000.

$$P = \frac{Annual\ Cost_{E} - Annual\ Cost_{R}}{Annual\ Cost_{E}} \times 100 \tag{1}$$

Where:

P is percent reduction

E is Existing fixtures

R is Recommended fixtures

Table 3: Data Collected from the BC Facility

Type of Area	Serial No.	Fixture Type	Fixture Details	Wattage	Quantity	Operating Hours per Year
Office	1	A	U Shaped Premium Lights Fluorescent Lamps FB40T12	40	74	6050
	2	B	4' F40T12	40	551	63000
	3	C	4' F32T8	32	6	3000
Miscellaneous	4	D	U Shaped Premium Lights Fluorescent Lamps FB40T12	40	14	3000
	5	E	135 watt Incandescent	135	7	6000
	6	F	4' F32T8	32	28	9000
Manufacturing	7	G	4' F40T12	40	224	30000
	8	H	8' F-96T12	60	228	24000
	9	I	4' F32T8	32	26	12000
Conference Rooms	10	J	U Shaped Premium Lights Fluorescent Lamps FB40T12	40	8	3000
	11	K	135 watt Incandescent	135	36	5000
	12	L	4' F40T12	40	18	3000

Table 4: Total Cost per Year for Each Existing Lighting Fixture at BC Facility

LIGHTING COST

Enter Cost of Energy per kWh = $ [0.095] /kWh

Fixture Type	A	B	C	D	E	F	G	H	I	J	K	L
Wattage	40	40	32	40	135	32	40	60	32	40	135	40
Quantity	74	551	6	14	7	28	224	228	26	8	36	18
Total Wattage	2960	22040	192	560	945	896	8960	13680	832	320	4860	720
Total Hours / Year	6050	63000	3000	3000	6000	9000	30000	24000	12000	3000	5000	3000
kWh / Year	17908	1388520	576	1680	5670	8064	268800	328320	9984	960	24300	2160
Cost / Year	1701.26	131909.40	54.72	159.60	538.65	766.08	25536.00	31190.40	948.48	91.20	2308.50	205.20

Total kW / Year =	57

Total kWh / Year =	2056942

Total Cost / Year = $	195409.49

Table 5: Energy Star Recommended Fixtures at the BC Facility

Type of Area	Serial No.	Fixture Type	Fixture Details	Wattage	Quantity	Operating Hours per Year
OFFICE	1	A	Energy star Fluorescent	30	74	6050
	2	B	Energy star Fluorescent	30	551	63000
	3	C	Energy star Fluorescent	23	6	3000
Miscellaneous	4	D	Energy star Fluorescent	30	14	3000
	5	E	Energy star Fluorescent	30	7	6000
	6	F	Energy star Fluorescent	23	28	9000
Manufacturing	7	G	Energy star Fluorescent	30	224	30000
	8	H	Energy star Fluorescent	36	228	24000
	9	I	Energy star Fluorescent	23	26	12000
Conference Rooms	10	J	Energy star Fluorescent	30	8	3000
	11	K	Energy star Fluorescent	30	36	5000
	12	L	Energy star Fluorescent	30	18	3000

Table 6: Total Cost per Year for Each Recommended Lighting Fixture at BC Facility

LIGHTING COST

Enter Cost of Energy per kWh = $ [0.095] /kWh

Fixture Type	A	B	C	D	E	F	G	H	I	J	K	L
Wattage	30	30	23	30	30	23	30	36	23	30	30	30
Quantity	74	551	6	14	7	28	224	228	26	8	36	18
Total Wattage	2220	16530	138	420	210	644	6720	8208	598	240	1080	540
Total Hours / Year	6050	63000	3000	3000	6000	9000	30000	24000	12000	3000	5000	3000
kWh / Year	13431	1041390	414	1260	1260	5796	201600	196992	7176	720	5400	1620
Cost / Year	1275.95	98932.05	39.33	119.70	119.70	550.62	19152.00	18714.24	681.72	68.40	513.00	153.90

Total kW / Year = 38

Total kWh / Year = 1477059

Total Cost / Year = $ 140320.61

Table 7: The Approximate Savings the BC Facility can Obtain per Year

Area	Fixture Type	Current Fixtures Annual Cost ($)	Savings Opportunity	Recommended Fixtures Annual Cost ($)	% Reduction in Cost	Annual Savings ($)
Office	A	1,701.26	Reduced Wattage	1,275.945	25.00	**425.32**
	B	131,909.40	Reduced Wattage	98,932.05	25.00	**32,977.35**
	C	54.72	Reduced Wattage	39.33	28.13	**15.39**
Miscellaneous	D	159.60	Reduced Wattage	119.7	25.00	**39.90**
	E	538.65	Reduced Wattage	119.7	77.78	**418.95**
	F	766.08	Reduced Wattage	550.62	28.13	**215.46**
Manufacturing	G	25,536.00	Reduced Wattage	19,152	25.00	**6,384.00**
	H	31,190.40	Reduced Wattage	18,714.24	40.00	**12,476.16**
	I	948.48	Reduced Wattage	681.72	28.13	**266.76**
Conference Rooms	J	91.20	Reduced Wattage	68.4	25.00	**22.80**
	K	2,308.50	Reduced Wattage	513	77.78	**1,795.50**
	L	205.20	Reduced Wattage	153.9	25.00	**51.30**
		195,409.49		**140,320.61**	**28.84**	**55,088.89**

P2 Calculations

The results obtained from using the P2 GHG calculator and P2 Cost Calculator are shown in Table **8** and Table **9**, respectively. Based on the recommendations made to the facility, there is a reduction of nearly 500 MTCO2e GHGs and savings of over $55,000 for the reduction of electricity usage from over 2 million kWh to less than 1.5 million kWh which is in accordance with the savings obtained from EAS (Table **7**).

Table 8: P2 GHG Calculator Output

	State of U.S.	Electricity Conserved (Input value)	Unit reported	Electricity Conserved (kWh)	GHG reduction (MTCO$_2$e)
Total Project	Ohio	579,883	kWh	579,883	492.823

Table 9: P2 Cost Calculator Output

	Electricity Conserved	Units	Unit cost ($/unit just selected)	kWh Reduced	Dollar Savings
Total Project	579,883	kWh	0.1	579,883	$55,088.89

Life Cycle Assessment (LCA) Findings

The LCA was carried out using the EIO-LCA model for both the current fixtures and recommended Energy Star fixtures. The "power generation and supply" sector was selected in the operational phase to run the EIO-LCA tool. Table **10** provides information regarding the GHGs generated, energy used, and hazardous waste produced for the cases of electricity consumed by the facility before and after implementing the recommendations. EIO-LCA calculated a reduction in greenhouse gas generation from 1,830 metric tons of CO$_2$e to 1,300 metric tons of CO$_2$e (Table **9**). There was also a reduction of 6.3 Terajoules (TJ) of energy, and 7,100 short tons (st) of hazardous waste generation (Table **9**).

Table 10: The Results of EIO-LCA for Both Existing and Recommended Fixtures per Year for the BC Facility

Factor	Type	
	Existing Fixture (0.195 Million Dollars)	**Energy Star Fixture (0.139 Million Dollars)**
Greenhouse gases (Total CO$_2$e)	1830	1300
Energy (terajoules)	21.8	15.5
Hazardous waste generation (short tons)	24500	17400

RECOMMENDATIONS

Lighting consumes 25–30% of energy in commercial buildings, and is a primary source of heat wastage. By upgrading the quality of lighting, the comfort level and productivity of an occupant can be increased considerably. This replacement of lighting helps in energy savings and improves the overall performance of buildings. By replacing the older lights with most recent and improved lights, a better control of lighting can be achieved. The appropriate positioning of light source not only reduces the wastage of light, but also improves the occupant's efficiency. Employing lighting controls is a good way to conserve energy. Major factors that affect conservation of energy include when and where the lights are to be used, length of time they will be on, and how bright they are. Timers, motion sensors and photo sensors can help monitor lighting fixtures. Maintaining and cleaning fixtures is one important factor in energy conservation, because dirty light fixtures can reduce the light output by up to 50%.

CONCLUSIONS

Based on assessments done using the EAS tool, the savings obtained from running a hypothetical scenario for BC Corporation was $55,000 while reducing electricity usage from 2,057,000 KWH to 1,477,000 KWH (Tables **4** and **6**). This was 29% of the original energy bills generated by the present lighting fixtures (Table **7**). Replacing the existing bulbs with low-wattage Energy Star bulbs was all it took to obtain these savings. From the P2 GHG calculator output it can be concluded that there is 490 MTCO$_2$e GHG reductions. It can be concluded from the EIO-LCA output that the emission of greenhouse gas has been reduced from 1830 metric tons CO$_2$e to 1300 metric tons CO$_2$e, which is approximately 29%. It was observed that there is also a 29% reduction in energy use and 29% reduction in hazardous waste generation when energy efficient lighting fixtures were adopted.

REFERENCES

[1]   U.S. Environmental Protection Agency (EPA), Pollution prevention programs, 2012. Available   from: http://www.epa.gov/ebtpages/pollpollutionpreventionprograms.html. [Accessed February 28, 2012.

[2]   Annual Energy Outlook 2010: With Projections to 2035, DOE/EIA-0383(2010), U.S. Energy Information Administration, Office of Integrated Analysis and Forecasting, U.S. Department of Energy; Washington, D.C, 2010.

[3]   The International Energy Outlook 2011, DOE/EIA-0484(2011), U.S. Energy Information Administration,, U.S. Department of Energy; Washington, D.C, 2011.

[4]   Eichholtz P, Kok N, Quigley JM. Doing Well By Doing Good? Green Office Buildings UC Berkeley: Center for the Study of Energy Market 2009.

[5]   Nevin R, Watson G. Evidence of Rational Market Valuations for Home Energy Efficiency. The Appraisal Journal1998; 66-4: 401–409.

[6]   Sartori I, Hestnes AG. Energy use in the life cycle of conventional and low-energy buildings: A review article. Energy and Buildings 2007; 39-3: 249–257.

[7]   US Department of Energy. Energy Plus Building Energy Simulation Software. Available from: http://apps1.eere.energy.gov/buildings/energyplus/. [Accessed February 28, 2012].

[8]   Kumar A, D'Souza F, Vashisth S, Software for Material Safety Data Sheets. Environ. Prog. 1996; 15- 2: S17-S23.

[9]   Kumar A, Thomas S. A Software Tool for Screening Analysis of Lean Practices. Environ. Prog. 2002; 21- 3: O12-O16.

[10]   Pendse R, Kumar A, Vijayan A. Development of a Spreadsheet to Determine Natural Ventilation Cooling Hours for a Commercial Hybrid HVAC System. Environ. Prog. 2005; 24-1: 16-23.

[11]   Vijayan A, Kumar A. Development of a Tool for Analyzing the Sustainability of Residential Buildings in Ohio. Environ. Pro. 2005; 24-3: 238-247.

[12]   Raman N, Vijayan A, Kumar A. Development of a Pollution Prevention Tool For Assessment of Hospital Waste. Environ. Pro. 2006; 25-2: 93-98.

[13]   Nimse P, Vijayan A, Kumar A, Varadarajan C. A Review of Green Product Databases, Environ. Pro. 2007; 26-2: 131-137.

[14]   Kadiyala A, Kumar A. Development of an Environmental Compliance Tool for Small Businesses. Environ. Pro. 2007; 26-4: 316-326.

[15]   Kadiyala A, Somuri D, Kumar A. Development of a Tool for the Assessment of Department Specific Hospital Waste. Environ. Pro. 2008; 27-4: 432-438.

[16]   Kadiyala A, Velagapudi S, Kumar A. Development of Chemical Identification Software for Multi-National Industries. Environ. Prog. Sustainable Energy 2009; 28-1:13-19.

[17]   Kadiyala A, Nerella VVK, Kumar A. Development of an Assessment Tool for Pollution Prevention and Energy Efficiency in Food Industry. Environ. Prog. Sustainable Energy. 2009; 28-3: 310- 315.

[18]   The University of Toledo (UT) PPIS Website. Available from: http://p2tools.utoledo.edu/. [Accessed February 28, 2012].

[19]   Kumar A, Rao HG, Vijayan A, Varadarajan C. Pollution prevention. Encyclopedia of Chemical Processing, NY 2006; pp. 2231-2246.

[20]   U.S. Environmental Protection Agency (EPA), Pollution prevention (P2). Available from: http://www.epa.gov/p2/pubs/resources/measurement.html. [Accessed February 28, 2012].

[21]   Carnegie Mellon Website. [EIOLCA, 2012]. Available from http://www.eiolca.net/index.html. [Accessed February 28, 2012].

[22]   Energy star website. Available from: http://www.energystar.gov/index.cfm?c=cfls.pr_cfls_lumens. [Accessed February 28, 2012].

[23]   Energy star website. Available from: http://www.energystar.gov/index.cfm?c=sb_guidebook.sb_guidebook_lighting. [Accessed February 28, 2012].

APPENDIX

Sample Energy Assessments [1]

Ashok Kumar[*], Alexander Spivak and Matthew Franchetti

University of Toledo, Toledo, OH, USA

Abstract: Energy assessment examples are presented and related to chapters in the eBook.

Keywords: Sustainability, energy efficiency, energy assessment, waste assessment, energy solutions, waste solutions, cost reduction, source reduction, recycling, reusing.

ASSESSMENT A: ENERGY AUDIT OF A UNIVERSITY (BY TECHSOLVE)

Related Chapters: Illumination, Boilers, Kitchen hoods

The university utilized some energy efficiency practices already in place, but there were no formal procedures or system in place. The critical action items defined for the university include:

- Awareness and Training.

- Targets and Key Performance Indicators.

- Metering and Monitoring.

- Understanding Performance and Opportunities.

- Planning.

Based on these recommended action items, an energy management plan for the university was developed, and a consulting firm conducted an energy audit to determine opportunities for energy savings. A management plan was customized for the university and covered the following topics:

- Leadership.

- Understanding.

- Planning.

- People.

- Financial Management.

- Supply Management.

- Operations and Maintenance.

- Plant Equipment.

- Reporting.

- Achievement.

The technical audit found several opportunities for cost savings shown in Table **A-1** including:

Table A-1: Energy Saving Opportunities for the University

Energy Opportunity	Estimated Annual Cost Savings	Estimated Savings (Kwh)	Estimated Greenhouse Gas Savings (tons)
Residence hall boiler plant – heat recovery, stack dampers	$5,500	78,000	49
Village Apartments – condensing boiler	$2,000	28,600	18
Modify kitchen hood make-up air	-	-	-
Condensing boilers	$10,800	154,000	97
High bay light fixture replacement	$8,400	120,000	76

ASSESSMENT B: ENERGY AUDIT OF A LOCAL GOVERNMENT (BY TECHSOLVE)

Related Chapters: Illumination, Windows, Doors, Insulation, Garage doors, Office equipment, HVAC

An energy assessment of the local government's safety center, public works, recreation center, and municipal building was conducted. The energy assessment included a questionnaire and walkthrough to identify energy saving opportunities and best practices. The following report provides a summary of the findings, recommendations, and resources to assist the local government's efforts towards energy savings.

Safety Center

The Safety Center houses the fire department, police department, non-emergency call center, emergency management services, and temporary jail. The building is approximately 40 years old, and is open 24 hours per day/seven days per week.

Energy Recommendations

- Caulk and seal remaining windows and doorways (use of an infrared detector to determine heat loss may be beneficial);

- Provide additional insulation to facility;

- Provide motion sensor light switches in areas not used on regular basis;

- Phase out existing appliances and office equipment and replace with Energy Star equipment;

- Complete re-lamp of building to T-8 or T-5 fluorescent lights and ballasts (see attached information);

- Garage doors that can be interlocked with heater to shut off heat when door is open.

Public Works

The Public Works Department is responsible for maintaining the streets, vehicles, drinking water plant, storm sewers, and signs. The department has an office and a garage that was converted from a construction company facility.

Some Recommendations for Energy Savings Include

- Use of motion sensors or timing switches to shut off lights.

- Employ practices prescribed by Motor Matters for determining motor efficiency, replacement, and replacement decision for drinking water operations. Information on Motor Matters can be found at http://www.motorsmatter.org/. TechSolve is hosting a one-day Motor Master DOE training session on April 24, 2007.

- City engineer should use electronic storage capabilities for documents and drawings to reduce paper consumption.

- Phase in use of Energy Star computer and office equipment, and appliances.

Recreation Center

The local government bought the facility for its resident's use, charging a small fee for residents and non-residents to use the facility. An outdoor pool and water park are currently being constructed. Inside various athletic activities are offered including pilates, spinning and yoga classes, a gym for miscellaneous adult and children activities, and a batting cage. A portion of the facility is being upgraded and reorganized to better meet the patron's needs.

Some Recommendations for Energy Savings Include

- Use motion sensors for areas to shut off lights not in use;

- Evaluate replacement of heating and cooling systems with new, energy efficient models and provide zoned heating to balance and direct heat and cooling where most needed—especially since building occupancy has changed from the original owner/design;

- Replace old fluorescent lights and ballasts with T-8 or T-5 lights, preferably the low mercury, green tip lights;

- Replace showers with water-saver shower heads;

- Develop and maintain outdoor pool pump system;

- Complete window replacement project.

Municipal Building

The municipal building was constructed in 1910 as a school building and has been remodeled to upgrade the building's appearance and energy efficiency. Some of the added features/energy-saving practices in this facility include:

- Replacement windows;

- Some Energy Star office equipment;

- Shut down of lights and computers at night;

- New air conditioning unit.

Some energy/waste saving opportunities found during the site walkthrough includes:

- Provide motion sensor switches for common areas such as bathrooms, conference rooms, *etc.*;

- Provide added insulation and ensure doors and windows are properly sealed;

- Designate one person to ensure lights, computers, fax and copying machines are turned off or are in sleep mode every evening;

- Double-side paper documents whenever possible; and

- Minimize use of personal fans, heaters, coffee pots, *etc.*;

- General Organization Energy Saving Tips;

- Specify Energy Star rated computer equipment, copiers, fax machines and appliances as part of daily procurement practices. Energy Star equipment can be found at www.energystar.gov;

- Employ computer recycling program to properly rid Safety Center of unusable computer equipment. One local recycler used by Hamilton County Environmental Services is Technology Recycling Group. TRG's contact information can be found at http://www.recyclegroup.net/;

- Vending machines are another opportunity. Attached is a fact sheet that provides energy saving tips. Another resource for vending machine energy savings is SnackMiser® and can be found at;

  o   http://www.austinenergy.com/Energy%20Efficiency/Programs/Energy%20Miser/snackMiser.htm.

  o   http://www.goodmart.com/products/511745.htm.

The Public Works Director was questioned as to the overall energy management practices for the local government. The following practices were in place:

- Management wants to reduce energy;

- Someone reviews the energy bills;

- General cost-saving ideas have been identified;

- Budget is prepared annually for each energy form;

- Back-up power has been assessed;

- Energy sources are turned off when not needed;

- List of energy saving projects;

- Automated controls to shut down equipment;

- Adequate maintenance staff;

- Preventive maintenance is conducted to reduce energy inefficiency;

- Monthly energy bills reviewed and recorded;

- Access to interval energy data; and

- Energy saving projects implemented in the last year.

Energy management practices that were not implemented:

- Energy policy in place;

- Energy on the agenda in manager's meetings;

- Assigned energy manager to oversee energy usage and projects;

- Provide general energy awareness;

- Share energy reduction opportunities with employees;

- Formal energy management training;

- Energy usage evaluated for trends;

- Cost savings for major equipment;

- Budgeted energy projects for next year;

- Energy reduction goals;

- Established metrics for facilities;

- Energy rates reviewed;

- Written instructions for energy equipment;

- Solicit energy-saving ideas from staff; and

- Verification of energy savings for projects.

Overall energy saving recommendations for local government:

1. Designate or hire an energy manager to oversee energy-saving projects, building operation, and new installations. The energy manager's role will be to:

a. Ensure that new energy-saving opportunities are implemented for existing equipment.

b. Watch for trends in energy usage to minimize excessive use.

c. Ensure that existing equipment is maintained and operated at maximum efficiency.

d. Oversee new installations and ensure that energy efficiency has been incorporated into the design.

e. Develop new ideas for energy savings through seeking innovative technologies and equipment.

2. Specify Energy Star equipment in purchasing department whenever feasible.

3. Evaluate the feasibility of green building designs and concepts for new and retrofitted installations.

Some information on green building design are:

- http://www.usgbc.org/.

- http://www.ciwmb.ca.gov/GreenBuilding/.

- http://www.greenbuildingsolutions.org/s_greenbuilding/index.asp.

- http://www.epa.gov/greenbuilding/.

- Participate in the local US Green Building Coalition workshops and events. See http://chapters.usgbc.org/cincinnati/.

  1. Have key personnel attend energy training workshops. Several workshops are planned through local organizations including motor systems and green building design. In addition, Energy Star offers on-line courses at http://www.energystar.gov/index.cfm?c=business.bus_internet_presentations.

2. Provide general awareness training to employees including placement of reminders at light switches, computers, *etc.* to turn off these items when not in use.

3. Conduct building commissioning for all new facilities and equipment. Verifying that the building or equipment meets the specified energy efficiency prior to project sign-off will help to ensure that the building/equipment will function as designed. Information on building commissioning can be found at:

- http://www.eere.energy.gov/buildings/info/operate/buildingcommissioning.html.

- http://www.eere.energy.gov/buildings/tech/commissioning/.

- http://www.aceee.org/buildings/projects/current/bld_cxny.htm.

- http://www.hoksustainabledesign.com/may02/Feature/commissioning.htm.

1. Incorporate energy management practices throughout the organization to ensure that energy savings are sustained. Development of an energy management plan will provide a framework for these energy management practices and will sustain energy savings.

2. Ensure community growth is appropriately planned to minimize waste and energy consumption. Information on community growth best practices can be found at http://www.glrppr.org/hubs/toc.cfm?hub=800&subsec=7&nav=7.

3. Investigate the use of EnFocus provided by Duke Energy for major building meters to review trends and peak loads.

See http://www.cinergy.com/enfocus/eissignup/eishomepage.asp for more information about this subscription offering from Duke.

1. Evaluate use of ODOD loan program to increase availability for energy project funds. See attached information about the loan program. The ODOD contact for the loan is Carolyn Seward (614) 466-4053.

2. Rebuild America offers an opportunity to connect with DOE professionals and other local governments to learn more about energy saving measures. For more information about this program, please see http://www.eere.energy.gov/buildings/program_areas/rebuild.html. The ODOD contact for this program is Dr. Manny Annunike, (614).

3. Energy Star has software that allows for organizations to assess their facilities and track trends. See http://www.energystar.gov/index.cfm?c=evaluate_performance.bus_p ortfoliomanager.

With these changes, it is expected that the local government will save approximately:

- $10,000 in annual energy costs;

- 143,000 Kwh of electricity; and

- 96 Tons greenhouse gas emissions.

ASSESSMENT C: ENERGY AUDIT OF AN OHIO SMALL AGRICULTURAL EQUIPMENT MANUFACTURER (BY TECHSOLVE/ UNIVERSITY OF DAYTON INDUSTRIAL ACHIEVEMENT CENTER)

Related Chapters: Illumination, Insulation, Garage doors, HVAC, Kitchen hoods, Compressors, Belt Conveyors

The company has participated in the ODOD Energy Achiever diagnostic. Based on this diagnostic, the company needs assistance in the areas of:

- Energy Saving Opportunities;

- Energy Supply;

- Awareness and Training;

- Reporting Systems;

- Energy Load Management.

The management plan for the company was completed. In addition, it was determined that energy savings could be achieved through:

- Improved heating of the manufacturing area with radiant heaters;

- Replacement of T-12 lighting with more efficient T-8 or T-5 lights;

- Improved building insulation;

- Providing vinyl flaps on garage door openings; and

- Replacement of old, inefficient saws, presses, *etc.*;

- Institute a preventative maintenance program to fix compressed air leaks;

- Reduce the pressure set point on the 30-hp compressor;

- Turn off the paint booth exhaust fan when the booth is not in use;

- Replace smooth V-belts with notched V-belts on all belt-driven applications.

The company has already implemented the compressed air repairs and shutting off the paint booth. Future plans include the replacement of lighting and adding insulation to improve overall building efficiency.

It is estimated the company will save more than **$15,000 per year** by implementing the no cost energy-saving practices.

Energy-Saving Measures

At a cost of 6 cents per kWh in Ohio for industrial sector, annual energy savings are (15,000/0.06)=250,000 kWh. At 1.34 lbs of CO_2 per kWh generation, annual CO_2 reductions are 335,000 lbs (167.5 tons).

ASSESSMENT D: ENERGY EFFICIENCY ASSESSMENT FOR A SNACK FOODS COMPANY (BY EISC)

Related Chapters: Illumination, Boilers, Water Heaters

A large snack food processor in Ohio, pursuing aggressive energy management goals, wanted an energy assessment performed to identify improvement and savings opportunities. Overall energy usage trends and usage distribution, boilers, plant lighting, water heater, and some electrical equipment were evaluated within the scope of this assessment. Comparison of gross finished products with energy usage indicates an increase in production efficiency from 2005 to 2006 from an energy usage perspective in Btu/lb and KWh/lb utilization. However, total costs have also increased for both gas and electricity from 2005 to 2006 due to increased utility rates. Therefore, energy input costs in cents per lb have both increase for gas and electricity. Two gas-fired boilers each rated at about 5 MM Btu/hr and operating 24 hrs/6 days a week provide steam for the renderers, water heater and some other uses in the plant. Gas usage by the boilers is estimated to be around 25-30% (about 45,000 MM Btu/year) of total usage. At an average recent cost of $9.50 per MM Btu, this usage costs about $428,000 per year. Through combustion testing, the estimated gross efficiency of Boiler 1 was around 70% at full load and 64% at idle. Boiler 4 seemed to be a bit higher at around 72%, and performing better at lower loads. Better oxygen control could improve Boiler 1's gross efficiency by roughly 5%, worth about **$10,000 per year**. For this, upgrades in boiler controls and possibly variable frequency drives for the blowers could be investigated further. Better recovery of stack heat for both boilers could improve performance by another 5% each, worth about **$10,000 per boiler yearly**. For this, a maintenance check on the extent of heat exchanger scaling should be followed by an investigation into boiler stack economizers. These measures could improve gross thermal performance of the boilers in the 80% range. Operation of multiple boilers, particularly to handle fluctuating loads and modulate between part-load and full load conditions, could be further improved with PLC & PC-based centralized control system. It is guessed that around 5% efficiency gains or more in the overall system could be achieved.

This could be worth around **$10,000 per year for each of the boilers**. Total plant lighting consumes around 550,000 kilowatt-hours annually, costing around $30,000 per year at the gross rate of 5.5 cents per kilowatt-hour. This is about 8% of the total annual electricity cost of around $350,000 and is low by industry standards. A lighting survey was conducted at the plant. While lighting intensities seem adequate in most general areas, quite low intensities were recorded in between equipments, equipment rears and corners. Many of these low intensity areas (particularly in the front and middle dryer rooms, meat dumper inspection stations, and separator room) are active processing and maintenance areas and improved lighting intensities in these areas could improve safety and operational convenience. A replacement of the plant metal halide fixtures with a good layout of T8 fluorescent fixtures with electronic ballasts together with some task lighting and lighting controls in specific areas of the plant can reduce lighting electricity consumption in the 40 - 50% range, simultaneously improving lighting quality approaching 50%. Direct savings in electricity would be in the **$12,000-15,000 per year**, not including any economic value for improved safety and productivity. This is approaching a 4% reduction in overall electricity consumption. Payback on capital is estimated in the 5 year range; this is somewhat higher than normal due to the expected higher cost of specialty fixtures needed in a food processing plant (enclosed, shielded, chemical resistant and suitable for wet location). Hot water usage is estimated at around 15,000 gallons/day, 5 days a week. Total water usage at the plant is estimated at 50,000 gallons/day (5-day average). The hot water needs consumes about 5000 MM Btu/year (about $50,000/year) and is about 12% of the boiler load (and only 3% of total gas usage). Overall hot water generation efficiency is expected at around 70% or below. Direct-fired condensing hot water heaters 95-99% efficient are available for sanitizing hot water needs at food processing plants. So, apparently a large efficiency approaching 30% can be gained in the hot water system with a new replacement. However, the low hot water consumption, and subsequent low natural gas utilization may yield a payback of 5 years or longer. 30% of estimated current consumption would be worth about **$15,000 per year** at a natural gas rate of $9.5/MMBtu, as shown in Table **D-1**.

Table D-1: Summary of Estimated Savings for Snacks Food Company

Energy Opportunity	Annual Savings	Estimated Savings	Estimated Greenhouse Gas Savings (tons)
2 Boilers @ $10K/year by oxygen control	$20K	2105 MM BTU ($20K/$9.5 per MM BTU)	123 tons (@117 lbs of CO_2 per MM BTU)
2 Boilers @ $10K/year by recovery of stack heat	$20K	2105 MM BTU ($20K/$9.5 per MM BTU)	123 tons (@117 lbs of CO_2 per MM BTU)
2 Boilers @ $10K/year by operation of multiple boilers	$20K	2105 MM BTU ($20K/$9.5 per MM BTU)	123 tons (@117 lbs of CO_2 per MM BTU)
Electricity	$12K - $15K	200,000 kWh ($12K/0.06 per kWh)	134 tons (@ 1.34 lbs of CO_2 per kWh generated)
Replacement of water heaters	$15K	1578 MM BTU ($15K/$9.5 per MM BTU)	92 tons (@117 lbs of CO_2 per MM BTU)

ASSESSMENT E: ENERGY EFFICIENCY ASSESSMENT FOR A LARGE FOOD SNACK COMPANY (BY EISC)

Related Chapters: Boilers, Water Heaters

An energy assessment was completed for a large snack food processor in Ohio to identify improvement and savings opportunities. Several efficiency improvement recommendations those are practical and realistic for implementation at this facility were provided. One of the potentially more immediate recommendations was modification to the ventilation system to reduce temperature stratification in the production room. Modifications to the ventilation system as describe in more detail below may potentially provide an attractive return on investment, ROI. This idca is clcarly onc that mcrits further engineering considerations because of the low capital costs and attractive ROI and estimated saving in the **$20,000 - $40,000 per year** range. Another recommendation that would potentially have the most significant impact of these energy efficiency ideas is using fryer stack waste heat to preheat the frying oil. Stack temperatures as measured on the roof indicate that there is a considerable amount of wasted energy. Collecting this waste stack heat and using it to preheat the frying oil would provide E2 savings all year, even in the hottest summertime days. The estimated savings are in the **$150,000 per year** range though further investigation would be needed into the available heat recovery equipment and installation. Examples of packaged heat recovery products specific to snack-food fryers include Heat & Control's Booster Heater (Model BH Series) for pre-heating cooking oil and their Heat Recovery System (Model HRS Series) for space and water heating or product drying. Several other manufacturers provide condensing economizers and heat recovery systems for use with boilers, turbines and food-processing fryers and ovens, like Combustion & Energy Systems' ConDex condensing economizer. Also available are integrated pollution control and heat recovery systems that incinerate oil, particulates and volatiles from fryer stacks and transfer the waste heat to the cooking oil (Heat & Control's KleenHeat Pollution Control Heat Exchanger, Model KHX Series).

Conversion to T8/T5 fluorescent lighting throughout the plant is already underway. The estimated impact of this change on electric usage is in the **$40,000 – $50,000 per year** range. The conversion from metal halide HID lighting to well implemented T8/T5 lighting in a food plant typically has a pay back ROI in the 3

- 5 year range due to the relatively higher cost of the enclosed and fixtures capable of handling the sanitizing chemicals and water sprays. Two A.O. Smith Dura-Max water heaters, rated at 1.81 MM Btu input & 80% nominal efficiency, provide hot water for sanitizing and other needs at the plant. Direct-fired condensing water heaters rated at or above 98% efficiency, like Kemco Systems' 99.7% efficient TE 100 Direct Contact Water Heater, are often used at food-processing plants for hot water supply. While overall hot water generation efficiency can be increased by as much as 20% (from 75% with existing heaters to 95%), the initial savings estimate of **$11,500 per year** for 24,000 gal/day hot water usage would not justify the capital expense of immediately replacing the existing systems with a direct-fired system. It was recommended that as the existing water heaters reach their end of life and need replacement or if plant expansion and modifications require a rework of the existing system, direct-fired water heating system should be given consideration and investigated (see Table **E-1**).

Table E-1: Summary of Estimated Savings for a Large Snacks Food Company

Energy Opportunity	Annual Savings	Estimated Savings	Estimated Greenhouse Gas Savings (tons)
Modification to Ventilation	$20K - $40K	333,333 kWh ($20K/$0.06 per kWh)	223tons (@1.34 lbs of CO_2 per kWh generated)
Collecting Waste Stack Heat	$150K	15,789 MM BTU ($150K/$9.5 per MM BTU)	923 tons (@117 lbs of CO_2 per MM BTU)
Electricity	$40K - $50K	666,666 kWh ($40K/$0.06 per kWh)	446 tons (@1.34 lbs of CO_2 per kWh generated)
Heater	$11,500	1210 MM BTU ($11,500/$9.5 per MM BTU)	71 tons (@117 lbs of CO_2 per MM BTU)

ASSESSMENT F: WASTE REDUCTION FOR AN AEROSPACE PARTS MANUFACTURER (BY TECHSOLVE)

Related Chapters: Office equipment, Documents

Background

A large aerospace supplier wanted to examine its wood and paper waste to determine if the waste could be reduced or recycled. The company employs 190 people and manufactures airframe structures and jet engine components for the military. The company had a recycling program already in place for scrap metal, oil and silver, but wanted to expand that program to include other waste streams. Through a pollution prevention assessment, the team would identify additional waste reduction and recycling opportunities, develop a comprehensive recycling program and assist the company's work towards its goal of becoming ISO 14001 certified.

Approach

Walkthrough the facility was aimed to determine:

- Types of wastes being generated;

- Method for storing and handling the wastes;

- Quantities of waste generation; and

- Processes generating the waste streams.

Based on this walkthrough, the team determined that the wood waste and office paper waste were the largest volume wastes, and that measures could be taken to reduce the waste. The team discussed ideas as to how to reduce these two waste streams and developed some options.

The wood waste being generated was from crates that were used to deliver parts, and pallets that were used for supply delivery. The company placed a separate trailer for collection of these wood wastes. To reduce the amount of wood waste, the team recommended that the company offer the crates back to the supplier or to find a local company that could use the crates. As a second alternative, the team recommended the company contract with a local recycling firm to prevent the wood waste from being disposed in the landfill.

Paper waste is being generated by the office staff during normal business activities and much of this paper is retention and eventual disposal of customer orders and drawings. Confidential documents are first destroyed in a paper shredder, and then disposed with the other municipal trash. The team recommended the following actions to reduce paper waste:

- Use electronic document storage whenever allowable;

- Double-side copies whenever possible;

- Scan and store documents on CDs or other electronic media; and

- Minimize paper magazines and have employees' access on-line media rather than ordering magazines and catalogs.

The team also provided the company with a free mercury pick-up option so that the company can remove the remaining mercury switches from its facility and reduce the possibility for an accidental release.

Table **F-1** shows details of the project:

Table F-1: Summary of Actual Savings for a Cincinnati Aerospace Parts Manufacturer

Project	Annual Cost Savings	Environmental Results	Status
Mercury Recovery	$1,500 – actual	30 grams of mercury	Fully implemented
Wood recycling	$1,800 – actual	480 yd3/yr	Fully implemented
Office recycling	TBD	37 tons /yr	In progress

ASSESSMENT G: ENERGY AUDIT FOR A FOOD MANUFACTURER (BY TECHSOLVE)

Related Chapters: Illumination, Boilers

Background

A local food manufacturing facility wanted to reduce its energy consumption by 5 percent. The manufacturing site employs approximately 60 people and makes food products for its restaurant chain.

Approach

The company's most viable option for addressing energy savings was the Ohio Department of Development's (ODOD) energy grant program. The program is geared towards companies that spend $350,000+ per year on energy, and it would enable the company to purchase new energy-efficient equipment with grant funds up to $50,000. The grant required the company to fulfill the following:

- Phase 1—Energy Diagnostic.

- Phase 2—Management plan development (third-party energy audit).

- Phase 3—Implementation (subsidy for equipment purchase).

Agreeing to the initial phase of the grant, the company participated in the energy diagnostic that was conducted by TechSolve on behalf of the ODOD. The commissary earned a —needs improvement ‖ rating, indicating an opportunity to move forward on Phases 2 and 3 of the grant. Projects identified in the third-party energy audit (Phase 2) are eligible for grant subsidy (phase 3) while funding lasts.

Results

Lighting

Most of the lighting in the commissary area is high bay, T12 fluorescent lights with low temperature ballast. A few of the areas have 400 W metal halide fixtures. The office lighting is mostly T12s, with a few areas that have been upgraded to T8. Numerical analysis is shown in Tables **G-1** and **G-2**.

Table G-1: Summary of Office Lighting Savings for a Food Manufacturer

Economic Summary	1 Fixture	100 Fixtures
Energy Usage, Initial, kwh	330	33000
Energy Usage,After upgrade, kwh	238	23800
Energy savings, %	28%	28%
Estimated Utility Cost Savings per year	$7.37	$737
Installed Cost (less rebate)	$86	$8,600
Simple Payback	11.6 years	11.6 years

Table G-2: Summary of Factory Lighting Savings for a Food Manufacturer

Economic Summary	1 Fixture	120 Fixtures
Energy Usage, Initial, kwh	1276	153120
Energy Usage,After upgrade, kwh	708	84960
Energy savings, %	44%	44%
Estimated Utility Cost Savings per year	$51.17	$6,140
Installed Cost (less rebate)	$300	$36,000
Simple Payback	5.8 years	5.8 years

Boiler Controls

Currently, the first boiler is set to maintain a certain pressure (85 psi) and the second boiler stages on when the pressure falls 15 psi in the header. There are a couple of energy-related issues with this scenario. If the pressure drop is a temporary one (which is what may happen when all process kettles fire on at once), by the time the second boiler is enabled and heats up the water from ambient temperature to the temperature needed for steam (usually about 15 minutes), there may no longer be a need for steam pressure and the second boiler goes into standby, without ever having contributed any steam to the system. A boiler staging control system would eliminate frequent second boiler cycling, and would automatically equalize run hours between them. See Table **G-3** for details.

In addition to the technical audit provided by the contractor, an energy management plan was developed to assist the company in sustaining energy savings through better energy practices such as:

- Strategic planning;

- Integrating energy into the business culture;

- Documenting all energy-related actions; and

- Establishing minimum criteria for addressing energy issues.

Table G-3: Summary of Boiler Savings for a Food Manufacturer

Initial usage to start 2nd boiler, 7x/week	5300.5
CCF utilized to start boiler 3x/week for 15 minutes with no steam added to header	2271.6
Gas usage after upgrade, 4 starts per week	3028.8
Percentage of usage saved, second boiler	42.8%
Cost of controller	$6,000
Utility cost savings per year	$2,681
Payback period	2.2 years

Calculation Summary

Total annual savings = $737 + $6,140.40 + $2,680.53 = $ 9,557.93

Energy saved in kWh/yr = [(0.28*33,000) + (0.44*153120)] = 76,612.8 kWh

Annual CO_2 reduction for savings of 76,612.8 kWh @1.34 lbs of CO_2 per kWh = 51 tons

Energy saved in MM BTU/yr = ($2,680.53/$9.5 per MM BTU) = 282.16 MM BTU

Annual CO_2 reduction for savings of 282.16 MM BTU @117 lbs of CO_2 per MM BTU = 16 tons

ASSESSMENT H: ENERGY AUDIT FOR A PRINTING COMPANY (BY TECHSOLVE)

Related Chapters: Illumination, Insulation, Kitchen hoods, Roofs, Heat Recovery

Background

A small printing company manufactures pressure sensitive, prime and bar code labels and tags, was faced with increasing energy costs. The company learned about the State's energy program called EnVinta and decided to participate to discover ways to reduce energy costs.

Approach

Required Phase 1 diagnostic was conducted using the EnVinta Achiever software. Based on the outcome of this diagnostic, it was recommended that the company focus on:

- Understanding saving opportunities.

- Metering and Reporting.

- Energy supply.

- Awareness and training.

- Targets and Performance Indicators.

In addition, it was also recommended that the company have a technical audit conducted to identify energy savings opportunities that could be funded through the State's energy program. The company agreed to proceed with Phase 2 part and have a technical audit conducted by a contractor and the energy management plan.

Results

Based on the technical review, the contractor recommended the company:

- Re-roof the facility with higher efficiency foam roofing materials;

- Re-lamp the facility with T-8 fluorescent lights; and

- Recover heat from the processing area.

ECM-1 - Replace the Shop Floor Roof with a Roof of R=12 or Greater

The building is masonry construction with a metal deck, ballasted EPDM (ethylene propylene diene monomer rubber) roof. The walls lack insulation, and the roof is minimally insulated (R=6). Because of the disruption to the building inherent with retrofitting insulation into the walls, this was not evaluated as a potential for energy savings. However, the roofs of both the shop floor, and the warehouse, are in need of replacement. It is recommended that both roofs be replaced with new roofs with increased insulation value. The new roofs are recommended to have an R value of 12 or greater to achieve the energy savings shown in Tables **H-1a** and **b**.

Table H-1a: Economic Summary for Printing Company by Replacing Shop Floor Roof with Roof of R=12 or Greater

Energy Usage, Initial, mmbtu/yr	467.6
Energy Usage, After upgrade, mmbtu/yr	257.2
Energy savings, %	45%
Estimated Utility Cost Savings	$3,694.96
Installed Cost	$59,345.00
Simple Payback	>20 years

Table H-1b: Economic Summary for Printing Company by Replacing Warehouse Roof with Roof of R=12 or Greater

Energy Usage, Initial, mmbtu/yr	120
Energy Usage, After upgrade, btu/yr	72
Energy savings, %	40%
Estimated Utility Cost Savings	$931.35
Installed Cost	$29,672
Simple Payback	>20 years

ECM- 2 – Install a Fan and Ductwork to Recover the Heat Rejected by the Air Compressor

There is potential for recovering the heat in the area of the air compressor. Another option would be to install a fan with a small amount of ductwork and an intake near the air compressor, to convey the warm air from the area around the compressor to either the second floor of the warehouse, or the first floor. Additionally, for occupant comfort in the warehouse, consideration should be given to adding an exhaust fan in the compressor area, to relieve heat buildup in the area in the summer. See Table **H-2** for details.

Table H-2: Economic Summary for Printing Company by Installing Fan and Ductwork

Energy Usage, Initial, ccf	2000
Energy Usage, After upgrade, ccf	1028
Energy savings, %	48%
Estimated Utility Cost Savings	$1186.62
Installed Cost	$10,000
Simple Payback	8.4 years

ECM-3– Install a Make-Up Air Unit with Gas Heat for the Printing Shop Area

Because of the volume of air exhausted from the space, doors are difficult to close and the strip curtain from the dock into the shop area was blowing inwards. This indicates that a high amount of infiltration of outside air is occurring. High infiltration increases cooling and heating load in the space. Typically, manufacturing spaces that exhaust a lot of air provide make-up air to the space. However, since this make-up air is not air conditioned, it would not lessen cooling load. The make-up air could be preheated before introducing it to the space. This may be a route to investigate after operating the space into the winter without the boiler. It may be possible to eliminate operation of the boiler entirely if heated, make-up air is provided to the printing shop area. If this option is pursued, the duct configuration could utilize pulling air from the area of the air compressor (to recover this waste heat) and using this air to preheat the outdoor air before it is introduced to the building. See Table **H-3** for details.

Table H-3: Economic Summary for Printing Company by Installing Make-up Air Unit

Energy Usage, Initial, CFH	4800
Energy Usage, After upgrade, CFH	2308
Energy savings, %	52%
Estimated Utility Cost Savings per year	$3043.84
Installed Cost	$25,000
Simple Payback	8.2 years

ECM-4– Upgrade Lighting in Print Shop Area

Lighting in the shop area is fluorescent strip lighting, with supplemental metal halides. The fluorescent strip lighting needs to use a certain type of bulb so that the color spectrum is appropriate for viewing printed material. The shop area has (40) 8 foot, 2 lamp fluorescent fixtures with what appeared to be magnetic ballasts. There are (16) metal halides, which have experienced considerable color change (indicating that the lamps are near the end of their service life). It is recommended that the lighting in the shop area be upgraded. The strip lighting should be replaced with T8 lamps of appropriate spectrum, with matched electronic ballasts. The metal halides should be replaced with high bay compact fluorescents. The upgrade of the high bay fixtures would provide the most utility cost savings, as shown in Table **H-4**.

Table H-4: Economic Summary for Printing Company by Upgrading Lighting

High Bay	1 Fixture	16 Fixtures
Energy Usage, Initial, kwh	2855	45680
Energy Usage, After upgrade, kwh	1583	25328
Energy savings, %	44%	44%
Estimated Utility Cost Savings per year	$120.82	$1933.12
Installed Cost	$390	$6240
Simple Payback	3.2 years	3.2 years
Strip Fluorescents, Shop	**1 Fixture**	**40 Fixtures**
Energy Usage, Initial, kwh	951	38068
Energy Usage, After upgrade, kwh	777	31080
Energy savings, %	18.3%	18.3%
Estimated Utility Cost Savings per year	$16.58	$663.20
Installed Cost	$95	$3800
Simple Payback	5.7 years	5.7 years

To date, the company has replaced its roof to the more efficient roof and is implementing the lighting retrofit. The energy management plan is designed to sustain energy savings through better energy practices including:

- Developing energy trends and reviewing demand and power factor;

- Providing general awareness training for its employees;

- Correcting energy waste as it is identified and identifying the resources that can assist with energy waste reductions; and

- Documenting efforts towards energy management.

Calculation Summary

Total annual savings = $ (3694.96 + 931.35 + 1,186.62 + 3,043.84 +1,933.12 +663.20) = $11,453.09

Annual savings in MM BTU = [(0.45*467.6) + (0.4*120) + (0.48*(2000 CCF*0.1 MM BTU/CCF)) + ((0.52*4800CFH*103BTU/hr*8765.8hr/yr)/106)] = 22,154.42 MM BTU

Annual CO_2 reduction for savings of 22,154.42 MM BTU @117 lbs of CO_2 per MM BTU = 1,296 tons

Annual savings in kWh = [(0.44*15680) + (0.18*38068)] = 26,951.44 kWh

Annual CO_2 reduction for savings of 26,951.44 kWh @1.34 lbs of CO_2 per kWh = 18 tons

ASSESSMENT I: ENERGY ASSESSMENT FOR A GARDEN (BY TECHSOLVE)

Related Chapters: Illumination, Boilers, HVAC

Background

A Garden has taken the initiative to go green and reduce its impact on the environment. The Garden has always been a leader in energy efficiency, and has Leadership in Energy and Environmental Design (LEED) certified facilities on site. However, the Garden's facilities are aging and some of the infrastructure is outdated and needs to be upgraded. The Garden chose to examine the State energy program because of its ability to address sustaining energy savings.

Approach

The initial Phase 1 energy diagnostic was conducted to determine the Garden's current energy management practices. According to the diagnostic results, the Garden needs to focus on the following elements to improve energy management:

1. Reporting, feedback, and control systems;

    a. Generate monthly reports depicting overall energy use per unit of activity (*e.g.*, kWh per area) and examine results where they show large cost or usage variance from target.

    b. Targets, key performance indicators (KPI), and motivation.

2. Set overall energy savings targets for reducing energy costs or improving energy efficiency based on benchmarking or an assessment of opportunities;

    c. Operating procedures.

3. Establish basic operating procedures/work instructions for all energy intensive processes and equipment;

    d. Purchasing procedures and alternative energy options.

4. Routinely review energy prices to determine if better rates can be attained;

    e. Metering and monitoring.

5. Regularly monitor the energy use of all major facilities/cost centers/energy-intensive end users;

6. Calibrate and regularly service energy metering/monitoring systems to ensure reliable data are available.

Consultant Company was hired to develop an energy management plan that will assist the Garden with their green initiatives and to sustain energy savings. The Garden was also in need of a technical audit to determine what projects would provide quick payback energy savings. A contractor was hired to conduct the energy assessment.

Results

Energy management plan was based on the results of the EnVinta diagnostic and the initiatives that Garden personnel were currently or planning to undertake to reduce waste and improve energy efficiency. The plan focused on several elements including:

- Developing a Green Team to help implement the plan elements, provide continual identification of energy and waste saving opportunities, and provide general awareness training;

- Documenting projects and other elements of the energy plan; and

- Providing a three-year strategic plan for project and element implementation.

ECM-1: Replace Primate Boilers with Modular Boilers

The primary boilers were installed in 1972. At the time the boiler system was designed, it was intended to put a roof over the outdoor yard and heat this area. This roof was never installed, but the extra capacity was installed into the boiler system, resulting in considerably more boiler capacity than is actually required for building usage.

Also, the boilers are 30 years old and nearing the end of their service life. Their estimated efficiency, according to a sales representative familiar with this type of equipment, is no greater than 80%, and probably less, due to their age. The boiler system consist of (2) fire-tube, 100 HP boilers, and (1) 500,000 btuh input boiler. The pumping system is primary/secondary. If the boiler system were replaced with (2) 2,000,000 btuh input condensing boilers, this would considerably improve system efficiency. The combustion efficiency would increase to 90 to97 percent, depending on water temperature, which would result in a corresponding decrease in gas usage. Overall pumping power would decrease, because the primary pumping power dramatically decreases when a condensing boiler is utilized. Full modulation of the boiler, rather than an oversized boiler with a 4 to 1 turndown, would result in boiler burner operation more closely in line with actual load, and would also decrease gas usage (see Table **I-1**).

Table I-1: Economic Summary for a Garden by Replacing Primate Boilers

Energy Usage, Initial, btu	4,311,953,605
Energy Usage, After upgrade, btu	3,599,013,917
Energy savings, %	16.5%
Estimated Utility Cost Savings per year	$9512.52
Installed Cost	$175,000
Simple Payback	18.3 years

The economics summary represents the savings due to boiler efficiency and pumping power only. Additional savings due to full modulation of an appropriately sized boiler, *vs.* an oversized boiler with limited turndown will also be realized.

ECM-2: Replace Autogate Boiler with a Modular Boiler

The Autogate boiler is original to the building, which dates from about 1968. It is a cast iron sectional boiler, and it has noticeable degradation of the tube sheet. It has past the end of its service life. Replacing this boiler would reduce gas usage in the Autogate. Gas usage for the last service year was 3495 CCF, or $4675. This is a fairly high number for a building of its size. See Table **I-2** for details.

Table I-2: Economic Summary for a Garden by Replacing Autogate Boiler

Energy Usage, Initial, btu	329,333,850
Energy Usage, After upgrade, btu	265,591,814
Energy savings, %	19.4%
Estimated Utility Cost Savings per year	$706.22
Installed Cost	$15,000
Simple Payback	>20 years

ECM-3

Continue to encourage personnel to raise their thermostat setting to 74-75 degrees in summer and 68 in winter. If thermostats are able to be fixed by HVAC maintenance, then set the thermostats at the above temperature.

Table **I-3** indicating energy saved from raising setpoint during cooling season is shown below:

Table I-3: Energy Saved for a Garden by Raising Setpoint

Dry Bulb Temperature	Btuh/1000 CFM Saved, Per Degree of Increase
72	--
73	2,700
74	2,657
75	3,000
Total saved (if thermostat is raised from 72 to 75)	8,357

For example, if the recently remodeled administration building, with 5 units at an average of 1,800 CFM were set up from 72 to 74 degrees during cooling season, this would result in an energy savings of approximately 48,213 btuh. At an average electrical use of 1.4 kwh/ton for 2500 cooling hours, this would result in a savings of 16,874 kWh or $1350 per year of electrical usage costs. This example is for one building—the savings would be greatly increased if the thermostat settings in all the buildings would be raised 2 or 3 degrees.

ECM-4: Lighting

Lighting systems in several buildings have been upgraded, or are under contract to be upgraded, to T8 with electronic ballasts, from T12 with magnetic ballasts. This upgrade reduces overall energy usage for lighting, reduces cooling load from lighting, and also, slightly improves power factor.

Duke Energy is currently offering small incentives to upgrade these fixtures. A comparison Table **I-4a** of lighting efficiencies and costs is provided below:

Table I-4a: Comparison of Lighting (T8 *vs.* T12) for a Garden

Fixture Type	Lumens/Fixture	Watts/Fixture	Cost per Fixture	Installation Costs	Rebate from Duke	Cost to Operate/Year
T12, magnetic ballast, 2 lamp,2x4	4558	172	$45 (replace lamp and ballast)	$35		$26.42
T8, electronic ballast, 2 lamp,2x4	5700	124	$55	$35	$4	$19.05

Continuously upgrade T12 lighting to T8 with electronic ballasts where applicable. The economic summary is shown in Table **I-4b**.

Table I-4b: Economic Summary for Upgrading Lighting at Garden

	1 Fixtures	100 Fixtures	500 Fixtures
Energy Usage, Initial, kwh	330	33000	165000
Energy Usage, After upgrade, kwh	238	23800	119000
Energy savings, %	28%	28%	28%
Estimated Utility Cost Savings per year	$7.37	$737.00	$3685.00
Installed Cost (less rebate)	$86.00	$8600	$43,000
Simple Payback	11.6 years	11.6 years	11.6 years

The Garden also has a wide variety of other light fixtures. Wherever it is possible, the light substitutions shown in Table **I-4c** should be made:

Table I-4c: Recommended Lighting Replacements for a Garden

Lighting Type	Higher Efficiency Substitution	Approximate Wattage Savings/Fixture
100 W incandescent	Compact fluorescent	50 watts or more
High bay metal halide	High bay compact fluorescent	200 watts

Christmas Lights

The Garden hosts an annual holiday event, which involves thousands of strands of Christmas lights. LED Christmas lights are beginning to be available, and advertise that they use one tenth of the power of traditional Christmas lights, which are incandescent. They are somewhat limited in design (not as many different options available) but more and more are becoming available.

The costs for these LED strings varied depending on design, but most were in the $8.00-$20.00 per strand, which is somewhat more expensive than a comparable incandescent strand. However, the cost is not prohibitively higher. The manufacturers claim that the bulbs last 200,000 hours, which would surely outlast the wiring. More research should be done in this area, such as evaluating several manufacturers for pricing and durability, and then, consideration should be given to replacing light strands as they fail with comparable LED lights if available.

Calculation Summary

Total annual savings = $ (9,512.52 + 706.22 +1,350 +3,685) = $15,253.74

Annual savings in MM BTU = $[(0.165*4311.95) + (0.194*329.33)]$ = 775.27 MM BTU

Annual CO_2 reduction for savings of 775.27 MM BTU @117 lbs of CO_2 per MM BTU = 45 tons

Annual savings in kWh = $[16,874 + (0.28*165,000)]$ = 63,074 kWh

Annual CO_2 reduction for savings of 63,074 kWh @1.34 lbs of CO_2 per kWh = 42 tons

ASSESSMENT J: ENERGY ASSESSMENT FOR A RESEARCH LABORATORY (BY TECHSOLVE)

Related chapters: Illumination, Motors, HVAC, Kitchen hoods, Roofs, Heat recovery

Background

Research Labs contacted Dr. Ashok Kumar at University of Toledo Civil Eng. for assistance. Supplemented by a US EPA Pollution Prevention grant through the University, this assessment was conducted by CIFT engineers Mr. Rick Mazur & Mr. Somik Ghose. Support was provided by maintenance and operations personnel at Research Labs. Akhil Kadiyala & Ravikanth Garimella, Graduate Assistants at the University also assisted in the survey and data analysis. Walk-through surveys & discussion sessions at site were conducted on May 15 and June 26, 2008.

Summary

The objective of this energy efficiency assessment (E2) was to analyze usage and compare usage metrics to industry-wide databases to identify efficiency improvement opportunities. Also, assess renewable energy opportunities.

Utility bills and other records from 2003 to 2008 have been analyzed for electricity and natural gas. Usage has increased in 2006 and 2007 following most recent expansions. Annual electricity usage is about 13.5 million kilowatt hours and costs $1.2 million at an average rate of 9 cents per kilowatt hour; about 50% of this cost is demand charges with average monthly billing demand of 3000 kVA at $18 per kVA. Annual natural gas usage is about 108,000 mcf (1000 ft^3) and cost $1 million at an average rate of $9.5 per mcf. The heating, ventilating and air-conditioning (HVAC) system is the major user of energy with ventilation and cooling consuming about 40% and 25% of total electricity respectively, and heating consuming over 90% of total gas. Water bills analyzed for 2008 indicate an average usage of 100 ccf (100 ft^3) or 75,000 gallons per day that would cost about $260,000 annually at $7.2 per ccf total for water & sewer.

Comparison to datasets from DOE & US EPA's Labs21 environmental performance program indicates a well-designed and optimized system at Research

Labs on an average. Usage metrics are in mid-range of Labs21 data and improvement opportunities exist at Research Labs to reduce consumption to best (low) ranges. However, Research Labs is subject to higher utility rates increasing total cost than average Labs21 data. Among featured energy efficiency technologies, Research Labs has already implemented many in the newer expansions like variable air volume (VAV) ventilation system and fume hoods; variable frequency drives (VFDs) for pump, fan and chiller motors; NEMA premium efficiency motors; exhaust air heat recovery; direct digital control (DDC) building automation system; power factor correction; efficient lighting with timers & photo-sensors; efficient cage-washers with counter-current rinse water recycling; and alkaline digester for waste management.

Further improvements could be possible with operational changes and best management practices, which are usually low-capital, fast-return opportunities but require a systems approach for implementation and performance verification. For example, manual or automatic night-time and unoccupied-time setbacks for HVAC & lighting; HAVC temperature, humidity or air flow set-point changes; dry pre-cleaning with squeezes before wet wash-down, *etc.* Larger improvements also could be possible with higher efficiency HVAC system and other equipment over the moderate efficient equipment at Research Labs; these are usually capital-intensive, longer-return opportunities, but with significantly higher savings numbers and incentives. Finally, renewable energy systems like geothermal heat pumps for HVAC application and others like wind turbines or photovoltaic panels could be integrated for sustainability. Various financial incentives are available for most E2 and renewable energy systems that help reduce installation capital cost and enhance the economics & return on investment.

Summary of Recommendations

- Investigate best available water-cooled chilled-water cooling system with closed-loop cooling towers, especially for new installation or replacement at end of older equipment life. Estimates show a **$17,000/yr** savings for a 200 ton unit over average efficient air-cooled water chillers at Research Labs. Consider lifecycle analysis and not first cost only.

- Consider best available roof top units, especially for new installation or replacement at end of older equipment life. A 10 ton high efficiency EER 12.5 (Energy Efficiency ratio) unit will save **$400/yr** in electric over older EER 9 units at Research Labs.

- Investigate further optimization of the HVAC system through E2 set-point changes in the building automation system. A 2% E2 improvement would be **$40,000/yr** savings.

- Investigate ventilation air change setback opportunities for night-time and unoccupied-times. A 12 to 4 air change per hour (ACH) reduction in lab spaces considering 1/3rd unoccupied time would save **$40,000/yr**.

- Investigate any remaining opportunities for variable frequency drives, NEMA premium efficiency motors, and waste heat recovery. Return on investments is often 1 year or less depending on applications.

- Investigate condensing feedwater economizer for a steam boiler, savings could be **$12,000/yr** for a 200 HP boiler. Investigate high efficiency condensing heating systems with 95-99% efficiency over moderate 80% efficient equipment at Research Labs.

- Investigate desiccant-based dehumidification system to replace part of cold-coil dehumidification that costs **$300,000**/yr.

- Investigate peak demand management opportunities with thermal storage and distributed generation. 800 kW of peak demand reduction during summer months could save **$75,000**/yr.

- Consider occupancy sensors & controls for lighting.

- Ensure high 0.97-0.99 power factor with correction capacitors.

- Consider R30 wall and R50 roof insulation for future expansions. Installation premium for higher insulation levels in new construction and major renovations usually pays back in 1 to 2 heating seasons.

- Investigate a vertical loop geothermal heat pump that operates as a dual heating & cooling system for part of total HVAC load.

- Consider financial incentives for E2 & renewable energy systems.

Calculation Summary

Total annual savings = $ (17,000 + 400 + 40,000 + 40,000 + 12,000 + 300,000 + 75,000) = $ 484,400

Annual savings in MM BTU = [(0.15(estimated % savings)*12,000 (Annual Savings)/9.5(Cost per MM BTU or mcf)] = 189.47 MM BTU

Annual CO_2 reduction for savings of 189.47 MM BTU @117 lbs of CO_2 per MM BTU = 11 tons

Annual savings in kWh = [(17,000 + 400 + 40,000 + 40,000 + 300,000 + 75,000)Total Savings/(0.09)Cost per kWh] = 5,248,888 kWh

Annual CO_2 reduction for savings of 5,248,888 kWh @1.34 lbs of CO_2 per kWh = 3516 tons

ASSESSMENT K: ENERGY ASSESSMENT FOR FOOD SAFETY SYSTEMS, OHIO (BY TECHSOLVE)

Related Chapters: Insulation, HVAC

Background

Food Safety Systems contacted Center for Innovative Food Technology (CIFT) for assistance. Supplemented by a US EPA Pollution Prevention grant through the University of Toledo Civil Engineering, this assessment was conducted by CIFT engineers Mr. Rick Mazur & Mr. Somik Ghose. Support was provided by several maintenance and operations personnel at Food Safety Systems. Walk-through surveys & discussion sessions at site were conducted on May 7 and June 18, 2008.

Summary

The objective of this energy efficiency (E2) assessment was to analyze energy usage and identify efficiency improvement opportunities at Food Safety Systems' Distribution Center. Food Safety Systems' Headquarter & Office has several energy efficient features in place including lighting and ventilation and adopted several green management practices including waste recycling.

Utility bills and other records from 2006 to 2008 have been analyzed for natural gas and electricity at the Distribution Center. Annual natural gas usage is about 2,300 mcf (1000 ft^3) and cost $27,500 at an average rate of $12 per mcf. The heating, ventilating and air-conditioning (HVAC) roof top packaged units consume almost 100% of total gas. Annual electricity usage is about 365,000 kilowatt hours and costs $26,000 at an average rate of 7 cents per kilowatt hour. The roof top units are also the major user of energy with summer cooling consuming about 35% of total electricity ($8,900/year) and winter ventilation another 20% of total electric ($5,300/year). Lighting accounts for 15% ($3,700/year) of total electric with the rest 30% assumed to be for processing and support.

Further improvements could be possible with insulation on the concrete block wall and air-circulation ceiling fans. An R10 insulation improvement on the current

10,000 ft² of R2 hollow-core concrete block wall will reduce heating energy cost by $10,000 per year; some additional savings from insulation would also occur in cooling mode during summer. Air-circulation fans have the potential of preventing air stratification in the high-bay areas, particularly during winter, and enhance year-round occupancy comfort in the working spaces. Regular inspection and maintenance of the roof top HVAC units will ensure efficient operation and prevent increase in energy usage. Finally, renewable energy systems like geothermal heat pumps for HVAC application and others like small wind turbines or small photovoltaic systems could be integrated for sustainability. Various financial incentives are available for most E2 and renewable energy systems that help reduce installation capital cost and enhance the economics & return on investment.

Summary of Recommendations

- Investigate R10 insulation system on the concrete block wall. The concrete hollow-core block wall is the single largest type of surface area with the least insulation, estimated at R2 (All the way to 20-feet high in the south and south-east and south-west corners and 7-feet high from the base in the north, west and the north-east corner). About 2" of suitable insulation, such as foam board could provide insulation of R10 or higher and an annual savings estimated at **$10,000**. With a material cost of $1/ft² or less for the insulation, a one year payback on materials is expected. Installation cost needs to be factored in for complete payback estimate.

- Examples for appropriate insulation on the concrete wall could be rigid foam panels or insulation batts with a stud wall setup or could be sprayed foam-in-place. Some information provided in appendix.

- Investigate insulation improvement for the current sheathed wall with only about 1.5 inch insulation. This insulation system is deteriorating with age and seems damaged at some locations.

- Alternatively for insulation, investigate exterior wall insulation for entire building. Several manufacturers provide this solution called Exterior Insulation & Finish System (EIFS).

- Investigate ceiling mounted large diameter (24" or higher) low-wattage slow-rpm fans at strategic locations in the high-bay area to facilitate better air circulation around partition walls and high storage racks. Heating and cooling energy will be reduced and occupancy comfort in working spaces will improve in both winter and summer. Annual total HVAC related cost is $41,200 and a 3% reduction would save **$1200/year**.

- Consider best available roof top units, especially for new installation or replacement at end of older equipment life. A 10 ton high efficiency EER 13.0 (Energy Efficiency ratio) unit will save over **$200/yr** in electric over older EER 9 units at Food Safety Systems.

- Implement a regular pre-season inspection and maintenance schedule for the roof top units. In particular, pay attention to the mechanical performance of the economizer. In general economizers can save up to 20% of the RTU's annual energy consumption. 20% of $6500/year in heating & cooling for each 20-ton unit like at Food Safety Systems will be **$1300/year**.

Calculation Summary

Total annual savings = $ (10,000 + 1,200 + 200 + 1,300) = $ 12,700

Annual savings in kWh = [(12,700) Total Savings/(0.07)Cost per kWh] = 181,428 kWh

Annual CO_2 reduction for savings of 181,428 kWh @1.34 lbs of CO_2 per kWh = 121 tons

ASSESSMENT L: ENERGY REDUCTION FOR AN OHIO VEGETABLE PROCESSOR (BY EISC)

Related Chapters: Boilers

An Ohio vegetable processor wants to optimize steam and reduce natural gas usage for pollution prevention. This assessment by EISC identified a **20 to 22%** improvement in steam usage and thereby an equivalent reduction in natural gas usage for steam boiler system of **5,000 MMBtu/year (at 20% reduction, resulting in avoided emissions of about 292 ton/year CO_2 and 750 lb/year of NO_X),** and recommended installation of a multiple-boiler sequencing controller. Working with the client, multiple price quotation from equipment vendors and suppliers have been procured and evaluated for a payback analysis, and implementation is targeted for the 2010 processing season.

ASSESSMENT M: ELECTRICAL ENERGY EFFICIENCY ASSESSMENT FOR A LARGE PLASTIC INJECTION MOLDING & EXTRUSION COMPANY (BY EISC)

Related Chapters: Illumination, Motors

Objective

Electrical energy efficiency assessment was completed for an electrical energy intensive Large Plastic Injection Molding & Extrusion Company in Ohio, with annual bills of $650,000 to $700,000.

Findings

Data on major electrical equipment was collected with the assistance of plant personnel through surveys and interviews. Information on individual operating hours, loading pattern and duty cycle of the appliances were also collected. Electrical consumption of 10.8 million KWH (at a cost of $750,000) is projected for a 12-month period.

Monthly electricity bills for the period Nov 2003 - Mar 2004 were analyzed. Detailed unit breakdown of the bills were obtained from The Ohio Edison Company and utilized in the analysis. The following are some of findings.

- Average peak demand per month is 2100 KVA (Nov 2003 – Mar 2004).

- Demand cost constitutes 50% of total cost of electricity. At $16.50 per KVA, annual demand cost totals to $415,800 ($16.50 * 2100 * 12).

- $375,000/yr would be a more conservative estimate.

- The KVA demand is not uniform over the production hours and peaks intermittently for short periods. The peaks occur in the first shift between 6 am and 2 pm. Some of these peaks seem to be due to early morning startup on Mondays.

- It seems possible to avoid these intermittent peaks for short periods with better equipment management. A 200 KVA (about 10 %)

reduction in the peak demand per month could save $39,600 annually (200 * $16.50 * 12) in electrical charges. A mere 5 % reduction or 100 KVA per month will result in savings of $19,800.

- On peak and off peak electrical usage are almost equal at 50 % of the total.

- The average on-peak load factor is above 0.8, which is considered good. Off peak average load factor of 0.53 brings down the total load factor to an average of 0.6.

- The average power factor for the period Nov 2003 – Mar 2004 is 0.98. Though 1 is an ideal value for the power factor, 0.98 is fair.

Shop Lighting

- The general shop lighting is with Sylvania M59 400 W metal halide lamps in Lithinoa Hi-Tek fixtures. In addition, there are some 1000 W metal halides in the warehouse.

- Fixture placement is irregular with little grid pattern and the fixtures are also at different heights from the floor.

- For task lighting in inspection tables, unloading tables and work benches, *etc.*, 4 ft 34W T12 (Sylvania F34/CW/SS) fluorescent lamps are used in 2 or 4 lamp fixtures. There are also some 8 ft 60W T12 (Sylvania F96T12/CW/SS) fluorescent lamps in 2 lamp fixtures.

- The foot-candles measured in general shop areas vary from 10 to 30 fc and task-lighted workbenches/inspection tables and unloading areas have a lighting level of 30 to 60 fc in general. A few Inspection tables are without task lighting with levels as low as 16 fc.

- The color inspection area between Injection Molding Machines 5 and 6, lit with 40W GE F40/SP65 fluorescent lamps have 250 fc lighting level, while the unloading table beside Machines 1 and 2 have a 160 fc lighting level. The Hot Room has a 8-12 fc lighting level and the warehouse has a 2-10 fc lighting level.

- In general, most parts of the shop are sufficiently lit for the required work. It is not 'essential' to increase the general lighting level in the shop. However, increasing the lighting level to a uniform 35-40 fc with proper task lighting on a 'need-to-light basis' would enhance the ambience greatly.

Table **M-1** shows the economics of replacing the 190 400W metal halides in the shop (including the Hot Room but excluding the Warehouse) with 4 and 6 lamp hi-bay fluorescent fixtures.

Table M-1: Replacement of Metal Halides with Multi-Lamp Hi-Bay Fluorescent Fixtures

I	Fixture Description		MH	Hi-Bay FL	Hi-Bay FL
II	Quantity of Fixtures		190	190	190
III	Lamps/Fixture		1	6	4
IV	Lamp Type		400 W MH	32 W T8	54 W T5 HO
V	Watts/Fixture (incld. Ballast)		450	224	234
VI	Cost/Fixture ($)		0	300	300
VII	Cost/Lamp ($)		40	5	10
VIII	Cost per kWh ($)		0.035	0.035	0.035
IX	Cost per kVA Demand ($)		16.500	16.500	16.500
X	Operating Hr/yr		6400	6400	6400
XI	Installation Time/Fixture (hr)		1	1	1
XII	Installation Labor ($/hr)		35	35	35
XIII	Re-lamp Time/Lamp (min)		15	15	15
XIV	Re-lamp Labor ($/hr)		25	25	25
XV	Expected Lamp Life (hr)		20000	20000	20000
XVI	Lamp Changes/yr	(II) * (III) / { (XV) / (X) }	61	365	243
XVII	Initial Fixture & Lamp Cost ($)	(II) * (VI) + (II) * (III) * (VII)	0	62700	64600
XVIII	Installation Cost ($)	(II) * (XI) * (XII)	0	6650	6650
XIX	Total Initial Cost ($)	(XVII) + (XVIII)	**0**	**69350**	**71250**

XX	Lamp Cost/yr ($)	(XVI) + (VII)	2432	1824	2432
XXI	Re-lamp Cost/yr ($)	(XVI) * (XIII)/60 * (XIV)	380	2280	1520
XXII	KW	(II) * (V) / 1000	86	43	44
XXIII	KVA	(XXII) / 0.98 ; PF=0.98	87	43	45
XXIV	kWh/yr	(II) * (V)/1000 * (X)	547200	272384	284544
XXV	Electricity Cost/yr ($)	(XXI) * (VIII)	19152	9533	9959
XXVI	Demand Charges/yr ($)	(XXIII) * (IX) * 12	17274	8599	8983
XXVII	**Total Annual Cost ($)**	(XX) + (XXI) + (XXV) + (XXVI)	**39238**	**22236**	**22894**
XXVIII	**Initial Cost Premium ($)**	-		**69350**	**71250**
XXIX	**Annual Savings ($)**	-		**17002**	**16345**
XXX	**Simple Payback (yr)**			**4.1**	**4.4**

Variable Frequency Drives on Injection Molding Pump Motors

The existing VFDs on the hydraulic pumps of the large injection molding machines have been found to be broken and taken off-line or not optimized for maximum electrical savings.

The following is the list of existing VFDs for the molding pump motors:

1. Injection Molder 4,150-hp pump motor, VFD not working;

    Electrical consumption without VFD = $23,100 per year.

    (assuming 5000 working hours, 20% duty @ 95% load and 80% duty @ 50% load)

    From studies done by Magnum AC Drive Systems at Carlisle Engineered Plants, an average of 40% savings is possible with properly programmed VFDs.

Savings Potential = \$9250

2.   Injection Molder 6, 150-hp pump motor, VFD not working;

Electrical consumption without VFD = \$23,100 per year.

(assuming 5000 working hours, 20% duty @ 95% load and 80% duty @ 50% load)

Assuming an average of 40% savings, **Savings Potential = \$9250**

3.   Injection Molder 5, 2 150-hp pump motor, 2 VFDs working;

VFD has been programmed to operate pump between 30 and 100% load. The load does not fall below 30%.

Electrical consumption with existing VFDs = \$16,600 each per year.

(assuming 5000 working hours, 10% Duty @ 95% Load, 30% Duty @ 50% Load, 60% Duty @ 30% Load)

Electrical Cost without VFDs = \$23,100 each per year (from 1 & 2 above).

Savings achieved = \$23,100 - \$16,600 = \$6,500 each (28%).

Assuming that a potential 40% savings is achievable with optimizing the operation of the VFDs,

Additional Potential Savings = \$9,250 - \$6,500 = \$2,750 each.

Potential Savings for both = \$5,500

4.   Injection Molder 6, 100-hp pump motor, VFD set at 75%;

Electrical consumption with existing VFD = \$19,582 per year.

(assuming 5000 working hours, 100% Duty Cycle @ 75% Load)

Assuming an average of 40% savings with proper programming,

Savings Potential = $7,800

5. Injection Molder 16, 100-hp pump motor, VFD working;

Electrical consumption with existing VFDs = $11,100 per year.

(assuming 5000 working hours, 10% Duty @ 95% Load, 30% Duty @ 50% Load, 60% Duty @ 30% Load)

Electrical Cost without VFDs = $15,400 per year.

(20% duty @ 95% load and 80% duty @ 50% load).

Savings achieved = $15,400 - $11,100 = $4,300 (28%)

Assuming that a potential 40% savings is achievable with optimizing the operation of the VFDs,

Additional Potential Savings = 40% * $15,400 - $4,300 = $1,860.

Total Potential Savings with repairing and optimizing all the 5 VFDs = $33,600 approx

ASSESSMENT N: ISO 14001 IMPLEMENTATION FOR A TRANSMISSION PARTS SUPPLIER (BY TECHSOLVE)

Related Chapters: Office equipment, HVAC

Company Profile

Transmission Parts Supplier requested assistance with the implementation of an ISO 14001 management system. The company has approximately 60 full-time personnel that manufacture transmission parts for Ford. Because the company has recently been designated as a Tier I supplier, Ford is requiring Transmission Parts Supplier to become ISO 14001 certified.

Approach

A cursory regulatory compliance audit was performed to determine what wastes should be prioritized for consideration as a significant aspects. Significant aspects were prioritized based on their severity of impact if released, the degree of impact, the quantity generated, and whether the waste is regulated. Based on these criteria, the following wastes or potential wastes were deemed to be significant:

- Electronics,

- Universal waste (fluorescent lights and batteries),

- Stormwater, and

- Ozone depleting substances.

Solutions

Because ISO 14001 requires that companies investigate possibilities for prevention of pollution whenever feasible, for the significant aspects identified based on available resources, the following solutions were suggested.

1. Electronics

On occasion, Company J will need to replace outdated computer hard drives, monitors, fax machines, and copy machines. Currently, Company J has outdated computer equipment that is unusable. Since electronic equipment can contain mercury and lead, electronic equipment was designated as a significant aspect. The amount of current electronic waste is unknown and the quantity that will be generated in the future cannot be estimated.

TechSolve located an electronics recycler that would be able to pick up their electronic waste and dismantle the parts and recycle them in an environmentally responsible manner.

2. Universal Waste

Fluorescent lights are occasionally changed out. The lights being used by the company are the higher mercury fluorescent lights. TechSolve is recommending the future use of low mercury lights. Also, Rumpke will recycle the fluorescent lights. No quantity of spent lamps was known at the time of this report.

Used batteries are generated from the forklifts and from the floor scrubber. A contractor replaces the forklift batteries. Company J, as part of the ISO program, will be requiring that the contractor fill out a Contractor Package that will ensure the contractor manages the battery waste properly. Floor scrubber batteries will be recycled when spent.

3. Storm Water

Company J is putting together a Spill Plan and Storm water Pollution Prevention Plan to be able to react to any spill or potential contamination of the ground or storm water. In addition, Company J will train its employees to be able to respond to spills in order to mitigate any environmental impact.

4. Ozone Depleting Substance

Company has a few air conditioning units containing Class II freons. Because these substances are regulated, TechSolve recommended that Company hire a certified contractor whenever work needs to be performed on these units. This will help prevent the unnecessary release of these substances to the air.

ASSESSMENT O: ENERGY ASSESSMENT FOR A PLASTIC MATERIALS AND RESIN MANUFACTURER (BY EWI)

Related Chapters: Illumination, Motors, Compressors

Objective

Energy assessment for Plastic Materials and Resin Manufacturer, a leading domestic producer of industrial resins and catalysts.

Findings

Detailed analysis of the different systems consuming electrical energy was conducted. The total annual power consumed from electrical equipments and lighting was determined from the electricity bills. The Total Annual Power Consumed for Electrical equipments and lighting are 5,477,952 kWh and 984,401 kWh respectively.

Utility analysis involved the analysis of the utility bills with the help of graphs and tables to identify trends, anomalies and billing errors. The following were identified from this analysis.

- Average price of electricity for this industry is 5.32 cents/kWh. The national average for industry is about 7.26 cents/kWh.

- Energy costs comprise about 0.8% of sales. The average for Company E is about 1.5%.

Average Avoided Cost

Demand = $11.60/kW

Energy = $0.033/kWh

Avoided cost represents the cost savings that can be achieved by modifying the usage by a given amount. It is based on the historical data provided by the City of Columbus outlining the plant's recent usage. The avoided cost was used to calculate savings in all of the recommendations provided below:

Recommendations

AR 1: Reduce Compressed Air Energy Requirement: Distribution System

Company E is presently using two (2) 75 HP screw compressors. During the assessment all air compressors were observed to be running. All were very hot to the touch, indicating full (or near full) load. Leaks and other compressed air distribution problems may be causing Capital Resin to create more compressed air than is needed. EWI estimated that this recommendation will save $10,393/year and have a payback of 8.7 months.

AR 2: Reduce Compressed Air Energy Requirement: Air Pressure

Once the air distribution system is buttoned up, it is always advantageous to look at the demand side. One easy way to capitalize on distribution improvements is for Company E to lower the air pressure for its compressors. It is estimated that this recommendation will save $3,489/year and have a payback of 7.2 months.

AR 3: Power Factor Correction on Critical Motors

During the assessment, it was discovered that several of the critical motors had extremely low power factors in one or two legs of the three-phase power. This may be due to a need for power correction at the motor, but more likely is due to a mixture of 'DELTA' and 'WYE' wiring configurations in the distribution system. Correcting this would require some time and the cost would be hard to estimate, so an estimate based on past experience is made. It is estimated that this recommendation will save $30,408/year and have a payback of 9.9 months or 1.3 years based on which option is required.

AR 4: Implement a Closed Loop Water Cooling System

Company E uses an average of 1,223 -1,000 cu. ft. of water a month in its cooling system. The water comes from the city of Columbus at a net rate of $10.98 per 1,000 cu. ft. If the company were to install a closed loop system, the water consumption would be limited to the make up water to replace evaporation and blow-off. Sewerage costs will be based only on the blow-off. With annual operating costs, it is estimated that this recommendation will save $63,936/year and have a payback of 3.2 years.

AR 5: Install Higher Efficiency Lighting

Company E consumes 984,401 kWh a year in lighting for a total cost of $32,485. The lighting is a mix of different types and lights, particularly in the production areas. EWI recommended a lighting replacement program that replaces approximately 460 T12 fluorescent lamps with energy efficient T8 lighting and electronic ballasts, plus replaces the mix of metal halide, mercury vapor, incandescent and high-pressure sodium lighting with energy efficient fluorescent lights and electronic ballasts. This will save $8,868 with a payback of 3.4 years.

AR 6: Implement Energy Management System

Company E can benefit from the current energy environment, which has given energy monitoring a vital management role. The most basic reason for monitoring energy use is to gather information about system operation, which can be used to:

Understand and control energy costs.

Improve facility operations and processes.

It is estimated that this recommendation will save $53,292/year and have a payback of 4.5 months.

Table **O-1** gives the summary of the recommendations.

Table O-1: Details on Recommendations

Annual Savings				Project Cost	Simple Payback
AR Resource CO_2 (lbs.)*			Dollars		
1	Electric Consumption Electric Demand Total	191,798 kWh 29.2 kW/mo.	441,135 lbs 441,135 lbs	$ 6,329 $ 4,078 $10,393	$7,500	8.7 months
2	Electric Consumption Electric Demand Total	64,386 kWh 9.8 kW/mo.	148,088 lbs. 148,088 lbs	$2,833 $ 1,364 $4,197	$2,080	7.2 months
3	Electric Consumption	621,960 kWh 71 kW/mo.	1,430,508 lbs. 1,430,508 lbs	$20,525 $ 9,883	$25,000	9.9 months

	Electric Demand Total			$30,408		
3A	Electric Consumption Electric Demand Total	621,960 kWh 71 kW/mo.	1,430,508 lbs. 1,430,508 lbs	$20,525 $ 9,883 $30,408	$40,625	1.3 years
4	Cost Reduction	5,868 kcf water/year	.	$63,133/year	$200,000	3.2 years
5	Electric Consumption Electric Demand Total	163,944 kWh 24.84 kW/	37,707 lbs. 37,707 lbs	$3,458 $5,410 $8,868	$30,360	3.4 years
6	Electric Consumption Electric Demand Total	1,159,332 kWh 108 kW/mo.	2,666,463 lbs. 2,666,463 lbs.	$38,258 $15,034 $53,292	$20,000	4.5 months
Total			4,733,677 lbs	$134,834		

*Assumes 2.3 lb. CO_2/kWh.

ASSESSMENT P: ENERGY ASSESSMENT FOR A COMPANY THAT MANUFACTURES FLEXIBLE PERMANENT MAGNET PRODUCTS (BY EWI)

Related Chapters: Motors, Compressors

Objective

Energy assessment for company that Manufactures Flexible Permanent Magnet Products, one of the nation's leading manufacturers of flexible magnets

Findings

Detailed analysis of the different systems consuming electrical energy was conducted. Readings were taken off the main process equipments that included Voltage, Amperage, Wattage, Volts-Ampere, Volts-Ampere Reactive, and Power Factor. Readings indicate a fair amount of low power factors among the critical equipment. This is a cause potential cause of excessive power draw that can cause the motors to run hotter that recommended and to fail early.

Utility analysis involved the analysis of the utility bills with the help of graphs and tables to identify trends, anomalies and billing errors. The following were identified from this analysis.

- Average price of electricity for this industry is 4.3 cents/kWh. The national average for industry is about 7.26 cents/kWh.

- Energy costs comprise about 1.45 % of sales. The average for industry is about 2-3%.

Average Avoided Cost

Demand = $1.19/kW

Energy = $0.0207/kWh

Avoided cost represents the cost savings you would see from modifying your usage by a given amount. It is based on the historical data provided by the City of Columbus outlining your recent usage. Avoided cost is used to calculate savings in all the recommendations.

Recommendation

AR 1: Reduce Compressed Air Energy Requirement: Distribution System

Company F is presently using a 50 HP Quincy and a 60HP Atlas Copco air compressors (screw). During the assessment the air compressors were observed to be running. They were both hot to the touch, indicating full (or near full) load. Leaks and other compressed air distribution problems may be causing creation of more compressed air than is needed. This recommendation will save an estimated $3,868/year and have a payback of 7.8 months.

AR 2: Reduce Compressed Air Energy Requirement: Air Pressure

Once the air distribution system is buttoned up, it is always advantageous to look at the demand side. One easy way to capitalize on distribution improvements is for Company F to lower the air pressure for its compressors. It is estimated that this recommendation will save $488/year and have a payback of 1.02 years.

AR 3: Power Factor Correction on Critical Motors

During the assessment, it was discovered that several of the critical motors had low power factors in one or two legs of the three-phase power. This may be due to a need for power correction at the motor or is due to a mixture of 'DELTA' and 'WYE' wiring configurations in the distribution system. Correcting this would require some time and the cost would be hard to estimate, so an estimate based on past experience is made. It is estimated that this recommendation will save $30,408/year and have a payback of 9.9 months or 1.3 years based on which option is required.

AR 4: Implement Energy Management System

Company F can benefit from the current energy environment, which has given energy monitoring a vital management role. The most basic reason for monitoring energy use is to gather information about system operation, which can be used to:

Understand and control energy costs.

Improve facility operations and processes.

It is estimated that this recommendation will save $21,480/year and have a payback of 11.2 months.

Table **P-1** gives the summary of the recommendations.

Table P-1: Details on Recommendations

AR		Annual Savings			Project Cost	Simple Payback
Resource		**CO$_2$ (lbs.)***		**Dollars**		
1	Electric Consumption Electric Demand Total	169,112 kWh 85.8 kW/mo.	388,958 lbs 388,958 lbs	$ 3,500 $ 368 $3,868	$2,500	7.8 months
2	Electric Consumption Electric Demand Total	46,647 kWh 7.1 kW/mo.	107,288 lbs. 107,288 lbs	$ 965 $ 101 $1,066	$500	5.6 months
3	Reactive Demand Reduction Motor Life Improvement	352 25%	N/A	$ 1,690 $ 13,815 $15,505	$20,000	1.3 years
3A	Reactive Demand Reduction Motor Life Improvement	352 25%	N/A	$ 1,690 $ 13,815 $15,505	$25,000	1.6 years
4	Electric Consumption Electric Demand Total	965,201 kWh 105 kW/mo.	2,219,962 lbs. 2,219,962 lbs.	$19,980 $ 1,500 $21,480	$20,000	11.2 months
Total			**2716208 lbs**	**$41,919**		

*Assumes 2.3 lb. CO$_2$/kWh.

ASSESSMENT Q: ALTERNATIVE WASTE DISPOSAL FOR AN AEROSPACE PARTS MANUFACTURER (BY TECHSOLVE)

Related Chapters: Solid Waste Management

Objective

To review machining company's waste to determine if there were alternative methods to disposing of the waste.

Background

The company is an aerospace parts manufacturer that employs approximately 400 workers. Parts are manufactured utilizing a variety of technologies including Multi-axis EDM drilling; 5-axis percussion drilling, fusion welding, and cutting and trepanning; and advanced brazing technologies.

Waste streams that the company wanted to find alternative disposal methods for are listed in the Table **Q-1** below.

Table Q-1: Alternative Disposal Methods for Waste Streams

Waste Stream	Description	Composition	Quantity Generated	Proposed Disposal Alternative
Activator waste	Waste dust from preparation of aluminide coatings for jet engine turbine components	Ammonium hydrogen bifluoride	25 kg/3 months	Recycler
Alumina grit	Aluminum oxide grit material used for surface preparation	99.43% Al2O3	Unknown	Recycler
Chromium and aluminum powder in vacuum bags	Chromium-aluminum chunklets to coat jet engine turbine parts—residual dust	Aluminum 42-46% Chromium 54-58%	Unknown	Recycler
Computer and electronic equipment	Out-of-date computers, monitors, printers, *etc.*	Can contain lead and mercury	Unknown- periodic upgrade of office equipment	Recycler
Copper wire	Wire ends from machining process are collected	Copper	2 gaylords per month	Recycler

The processes that generated the above waste were reviewed to determine if there were any pollution prevention methods to reduce the waste at the source. The processes did not lend to changes to reduce these wastes, so the next step was to determine if there were recyclers able to recycle these materials. The dusts that were generated in the first three waste streams listed were from new processes so little waste had been generated at the time of this report. Metal recyclers were identified for these materials and the company is currently investigating these options.

ASSESSMENT R: ENERGY ASSESSMENT FOR A STAMPING & WELDING COMPANY (BY EISC & UNIVERSITY OF TOLEDO)

Related chapters: Illumination, Motors, HVAC, Plant machinery

Background

A facility-wide energy audit was conducted s for a small metal auto-parts manufacturer. This was the first step to identify savings and efficiency improvement opportunities. Stamping and welding are the two major processes at this operation with annual electrical charges of about $110,000.

The audit revealed opportunities for efficiency improvement in several areas including stamping press motors, air compressors, lighting and natural gas usage for comfort heating. Findings are summarized below.

Procedure

Motors

A detailed analysis comparing premium efficiency motors to rewound average efficiency motors was performed using DOE's MotorMaster. The general recommendation was to replace, instead of rewinding, the existing A.E. motors with P.E. motors on failure. 5 to 20 hp motors would provide a less than 2 year payback, larger motors would provide a less than 3 year payback (see Table **R-1**).

Table R-1: Motor Analysis

HP/RPM/V	Efficiency (AE Motor)	Efficiency (PE Motor)	Motor Premium	Energy Savings	CO_2 Savings (lbs.)*	Annual Savings	Simple Payback
5/1800/230	83.0 %	90.2 %	$171	811 kWh		$122	1.4 yr
10/1800/230	85.7 %	91.8 %	$307	1316 kWh		$197	1.6 yr
15/1800/230	86.5 %	92.7 %	$510	1959 kWh		$294	1.7 yr
20/1800/230	88.3 %	93.3 %	$527	2029 kWh		$304	1.7 yr
25/1800/230	88.9 %	93.8 %	$891	2470 kWh		$370	2.4 yr
30/1800/230	89.2 %	93.9 %	$982	2815 kWh		$422	2.3 yr
40/1800/230	89.4 %	94.6 %	$1528	4081 kWh		$612	2.5 yr
50/1800/230	90.6 %	94.9 %	$1693	4145 kWh		$622	2.7 yr
Total				19626	35326.8	$2,943	

- 3000 hr/yr, 75% Duty @ Full Load, $0.15/kWh.

- Motor efficiencies and average motor prices from MotorMaster database, 25% discount on list prices.

- 1% efficiency loss due to rewinding.

- Motor Premium is the difference between discounted PE motor cost and AE motor rewinding cost.

Lighting

Existing Metal Halide Lighting (as shown in Table **R-2**)

Sylvania 400 Watt Clear MetalArc Bulb.

50 Watt Ballast + Lithonia Fixtures.

Initial Lumens: 32,000 Mean Lumens: 20,500 Rated Life: 20,000 hr CRI: 65 CCT: 4000 K Warm-up Time: 2-4 min Hot Restrike Time: 7-12 min

Table R-2: Lighting Analysis

Location	Number of 450W Fixtures	Hours/year	$/year @ 15 ¢/kWh
Stamping Room	36	4000	$9,720
Brake Press Room 1	8	4000	$2,160
Brake Press Room 2	18	4000	$4,860
Compressor & Press Room	11	4000	$2,970
Shipping Area	15	4000	$4,050
Sub-Total	88	$23,760	
Weldshop	8	2000	$1,080
Laser Building	24	2000	$3,240
Sub-Total	32	$4,320	
Total	120	$28,080	
Total @ 90%		$25,270	

- 2 shifts, 16 hr/day, 5 day/wk, 50 wk/yr = 4000 hr.

- 1 shift, 8 hr/day, 5 day/wk, 50 wk/yr = 2000 hr.

Multi-lamp Fluorescent Fixtures

- 4-lamp 32W 4' T8 with one 4-lamp Ballast: 160 W, 11,000 lumens.

- 6-lamp 32W 4' T8 with one 4-lamp and one 2-lamp Ballast: 225 W, 17,000 lumens.

- 2-lamp 54W 4' T5HO with one 2-lamp Ballast: 120 W, 10,000 lumens.

- 4-lamp 54W 4' T5HO with one 4-lamp Ballast: 235 W, 19,000 lumens.

- 6-lamp 54W 4' T5HO with one 4-lamp and one 2-lamp Ballast: 350 W, 31,000 lumens.

T8 and T5 Fluorescent Lamps

- 80+ CRI, 95% Lumen Maintenance, Instant On-Off, Electronic Ballast, Compatible with occupancy sensors,

Estimate of Energy Savings

- A 6-lamp 225 watt T8 or a 4-lamp 235 watt T5HO fixture can replace a 400 watt metal halide fixture on a one-to-one basis (see Table **R-3**).

Table R-3: Energy Saving Estimate for Lamps

Fixture	Number	Hr/yr	Energy Savings (kWh)	CO_2 Savings (lbs.)*	Savings ($/year @ 15 ¢/kWh)	Cost @ $150 / fixture	Payback
225 W 6-lamp T8	88	4000	79200		$13,200	$11,880	1.1 yr
225 W 6-lamp T8	32	2000	14400		$4,800	$2,160	2.2 yr
Total			**93600**	**168480**	**$14,040**		

- Further savings can be achieved by strategic use of 2-lamp and 4-lamp T8 fixtures along with 6-lamp T8 fixtures

COMPRESSED AIR

Coolest Available Air for Compressor Intake

Whenever feasible, the intake for an air compressor should be run to the outside of the building, preferably on the north or coolest side. Since the average outdoor temperature is usually well below that in the compressor room, it normally pays to take in cool air from outdoors. The energy savings potential in lowering the air intake temperature results from the fact that colder air is more dense, and therefore a given pressure increase may be obtained with less reduction of volume of the air. This in turn means that the compressor does not need to work as hard to obtain the desired pressure.

Potential Savings

(Rutgers Office of Industrial Productivity and Energy Assessment)

1. Intake Air Temperature = 85 F.

2. Available Cool Air Temperature = 50 F.

   Fractional Reduction of Compressor Work = (85 - 50) / (85 + 460) = 6.4 %.

3. Annual Electricity Usage by Hill Manufacturing's Kaeser 91 Air Compressor.

   (75 hp, 4000 hr/yr, 75% duty, $0.15/kWh) = 168,000 kWh = $25,000 ($21,500 @ 85% of 168,000 kWh).

4. Potential Savings @ 6.4 % = $1600 ($1370 @ 85%).

Setback Thermostats for Laser Building

Summary

Annual Gas Cost (approximate): $7,000 (8000 ccf).

Potential Savings with Setback Thermostats: $2,250 (2400 ccf).

Capital & Installation Cost (4 to 6 Digital Thermostats): $1,000 to $1,500.

Gas Usage Analysis

2004 Jan-Dec Total Gas Cost = $7182.16

Usage = 8166 ccf

Average Gas Cost = $0.833 per ccf (2003-04 winter), $0.945 per ccf (2004-05 winter)

Climate Data

Toledo Ohio Annual Heating Degree Days = 6579

http://www.climate-zone.com/climate/united-states/ohio/toledo/

Days/year with Average Temperature below 65 F = 220 days

(Assumed from Climate Data, October – April, 7+ months)

Calculations

(Rutgers Office of Industrial Productivity and Energy Assessment)

1. Percent of weekly time when Laser Building not in operation assuming 50 hr/wk operation = (168 – 50) / 168 = 70 %.

2. Average temperature difference between indoors and outdoors in winter with 70 F indoor temperature = Temp. Indoors – {65 – Heating Degree Days / Days below 65 F} = 70 F – {65 – 6579 / 220} = 35 F.

3. Energy Savings in ccf assuming off hours setback to (70-15) = 55 F = Off Hours Temp. Reduction * % Off Hours * Annual ccf Usage / In-Out Temp. Difference = 15 F * 70 % * 8000 ccf / 35 F = 2400 ccf.

4. Annual Savings = 2400 ccf * $0.945 / ccf = $2,270.

5. Approx. Cost of Digital Seven Day Programmable Setback Thermostat = $200.

(Approx. Cost of Analog Thermostat = $100)

Installation Cost = $50 per thermostat)

Required Digital Thermostats = 4 to 6

Implementation Cost = $1000 to $1500

Simple Payback = 6 to 8 months (One winter season).

ASSESSMENT S: ENERGY ASSISTANCE FOR AN ALUMINUM CASTER (BY TECHSOLVE)

Related chapters: Motors, Compressors

The company provides aluminum and zinc cast parts to the automotive and non-automotive industry. One cost-savings measure they chose to pursue is energy efficiency. The company had out-dated equipment. Energy diagnostic under the Ohio Department of Development's (ODOD) Envinta program was conducted. (See http://www.odod.state.oh.us/cdd/oee/EnVinta.htm). As a result of this energy management diagnostic, the company was assisted to obtain a new melter and holding furnace.

The Envinta One-2-Five® energy diagnostic was conducted and identified several areas for improvement in energy management practices. The critical items identified during the diagnostic session were:

1. Metering and monitoring of energy-intensive equipment,

2. Operating procedures that contain the best energy efficiency practices for the process/equipment,

3. Maintenance procedures that contain the best energy efficiency practices for the process/equipment,

4. Reporting, feedback and control systems to identify and correct energy variances, and

5. Targets, performance indicators, and motivation which include establishment of goals and recognizing achievement.

A management plan to address these immediate needs was developed at the time a technical audit was performed to identify energy-saving projects and assign a strategic plan for implementation. As a result of this audit, the following projects were identified:

•   New melting furnace at less than 1/2 the rated energy consumption.

- Ladle pre-heater rated at less than 1/2 the current energy consumption.

- Energy-efficient motor installation.

- Compressed air system upgrade to eliminate one compressor and add storage capacity.

Implementation of a new stack melter and crucible holding furnace was recommended based on the technical audit. Potential savings of $235,000 are estimated from these two projects with a payback of 1.66 years after the State grant of $50,000. The existing melter has a measured energy usage of 2,419 Btu/lb. The new melter is rated at 1,100 Btu/lb total with the projected resultant savings of 55% or 22,800 MMBtu per year natural gas. The company has installed the new furnace and is currently verifying the cost and energy savings.

ASSESSMENT T: ENERGY ASSISTANCE FOR A COINING COMPANY (BY TECHSOLVE)

Related Chapters: Boilers, HVAC, Compressors

The company produces coins for specialty applications in a variety of metal alloys. In a campaign to reduce operating costs, the company decided to pursue energy cost reduction. Energy diagnostic was conducted under the Envinta program. Following the diagnostic, the two consultants identified the following energy saving measures:

1. Optimizing the compressed air system.

2. Adding programmable thermostats to the HVAC system for night set back.

3. Installing a high efficiency hot water boiler for heating.

By implementing these projects, the company is expected to reduce their electric consumption by approximately 50,000 kwh per year resulting in 90,000 lbs/year of reduction in CO_2 emissions. In addition to these energy saving projects, a management plan was developed to sustain energy awareness and savings activities. Through these efforts, the company is expected to save between $18 and $27 K or between 17 and 25 percent of their energy expenditure. The company is preparing to implement these changes.

The management plan developed based on the energy management diagnostic focused on:

- Accountabilities and responsibilities;

- Energy supply contracts;

- Awareness and training;

- Reporting systems; and

- Energy load management.

An energy management plan was developed to address these key energy management elements including a records maintenance matrix, equipment register, and a training matrix. Two maintenance personnel participated in a Compressed Air Challenge workshop that was sponsored by the US Department of Energy as part of the training plan established for the company. The company has since purchased $25K in new energy efficient motors to reduce energy costs associated with production equipment.

ASSESSMENT U: ENERGY ASSISTANCE FOR A HEAT TREATER (BY TECHSOLVE)

Related Chapters: Motors, Compressors

An energy diagnostic under the Envinta program was conducted. Following the diagnostic, an energy management plan developed and a technical audit conducted.

The company underwent the Envinta diagnostic and energy management plan development. This company had approximately $700,000 in annual energy costs. Projected savings of $40,000 to $80,000 were identified in these areas:

1. Develop and implement an electric demand management strategy.

2. Relocate compressors.

3. Install variable speed drives for oil agitators.

The company is projected to save approximately 98,000 kwh/yr (176,400 lbs/year reduction in CO_2 emissions) in electrical consumption by implementing these changes. The company is currently preparing to make these changes. In addition, the company is upgrading its office areas and plans to incorporate energy efficient lighting and HVAC system as part of this remodeling project.

The main energy management practice improvement elements identified during the Envinta diagnostic include:

- Reporting, feedback and control systems;

- Understanding energy performance opportunities (energy audit);

- Targets and key performance indicators;

- Demonstrated corporate commitment; and

- Metering and monitoring.

Improvements in the area of peak demand control, performance measures, and added monitoring capabilities are being considered to control energy use and to avoid unnecessary energy costs.

ASSESSMENT V: WASTE REDUCTION AT THE TURKEY FACTORY (BY EISC)

Related Chapters: Illumination, HVAC, Compressors, Water heaters

An energy assessment was conducted at a turkey slaughterhouse and processing plant. Energy usage trends, peak demand, refrigeration system & freezer insulation, plant lighting and water heating were evaluated to identify improvement opportunities.

UT students received data about the load, suction, discharge, oil, inlet temperature, discharge temp, suction temp and compressed air for various compressors. The data were entered into an excel sheet for analysis. The missing data points were linearly interpolate. Individual regression of load *vs.* various parameters was done to obtain a relation between them. But since the relation for all the parameters could not be obtained a multiple regression analysis was performed. The graphs showing the variations of each parameter for a particular month were drawn. Motor data are being analyzed.

Utility System Peak Demand

The plant is subject to a fairly complex peak demand billing schedule that includes the absolute peak demand at the plant as well as the plant's contribution to the utility's total system demand during specific monitored periods beyond the control of the plant. Attention is paid by plant personnel to track these system peak monitoring periods and attempts are made to reduce load at the plant to keep peak system demand low. However, this is done mostly intuitively rather that with a fool-proof systematic approach. A very high system peak got recorded in mid 2005 and was applied in billing for each month over Aug 2005 – Jul 2006. This increased demand charges by an average of $20,000 per month for the entire 12-month period, costing the company an extra $240,000 over the period. A new lower system peak recorded in July 2006 will reduce the demand charges substantially this year. While peak demand management is not directly related to energy efficiency, it has major financial implications for this plant.

Refrigeration System and Controls

Refrigeration systems and controls is potentially the most significant area for improved energy efficiency at this plant. A review of potential alternate refrigeration system controls was beyond the scope of this project but several ideas that could provide additional energy efficiencies were identified. The vapor compression system with screw compressors using ammonia refrigerant currently used at this plant is ranked as the most efficient for this type of food processing application. The existing refrigeration system is large and consists of multiple compressors, condensers, evaporators and refrigeration zones. There is no centralized control system and many of the components operate independent of each other, reducing overall system efficiency. There is some limited compressor sequencing in response to fluctuating plant load. Some manually recorded data on compressor operation was analyzed for several months in 2006. The data indicates that several motors operate at what appears to be an excessive light load compared to the overall demand for refrigeration. The complex interactions between the various pieces of equipment are not simple to track, correlate and understand with respect to maintaining an energy efficient operation. Some strategy for a better system monitoring and understanding was outlined for the plant.

1. The refrigerant discharge pressure is recognized as an important parameter to monitor in order to operate at high energy efficiency. It can also be used to indicate oil separator blockage, condenser performance and presence of noncondensibles.

2. Floating compressor pressure head (discharge pressure) control strategy is becoming more widely used today instead of the fixed head pressure system. Floating head pressure reduces the compressor related energy required as well as providing other system advantages and reduces the overall energy used if condenser fans and other equipment is not operated excessively to compensate for an excessively low head pressure.

3. Power used by each individual compressor motor.

4. Refrigerant leak detection monitoring.

5. Air temperature in refrigerated spaces.

6. Refrigerant temperature at the compressor discharge.

7. Temperature of the suction gas to the low stage compressor.

8. Liquid level in the high pressure receiver.

9. Status of the hot gas defrost operation (not sure this is applicable at this plant).

10. Water that enters the refrigerant piping from air entering the system during periodic routine maintenance on the piping, valves, *etc.* can reduce overall system efficiency by several percentage points. At a specific required evaporator temperature, the addition of water to the ammonia refrigerant will require a lower suction pressure. One source (Hansen Technologies) estimates that the extra energy required by a system with 10% water in the ammonia is 7.6%.

11. A plugged suction line filter and a low refrigerant charge are two very common problems that reduce the system performance. Monitoring of these and other parameters is recommended to maintain system performance.

Lighting

A walk-through lighting level survey was conducted at the plant with a light meter to measure lighting intensity in foot-candles at various locations in the processing areas, adjoining storage and shipping areas, and maintenance offices. Metal halide HID lighting is used throughout the plant. Although MH lighting has made advances in the past years, fluorescent technology has advanced further than MH for this type of application. It was recommended that the lighting be changed to multi lamp fluorescent fixtures (low temperature, wet location, shielded fixtures for food processing application). Changing the fixtures to multi lamp fluorescent lighting would:

*   Improve the lighting level as measured by foot-candles on the shop floor work area.

*   Improve the color rendition index, CRI, which improves the actual and perceived quality of the lighting in the work area.

- Reduce the heat load in the refrigerated and cooled area of the plant by replacing a high wattage HID fixture with lower wattage fluorescent lights.

- Reduce the overall energy used by the lighting system including reducing the peak KW Demand.

- Reduce the overall cost of electricity on a monthly bill basis.

The recommended next steps are:

- Identify a fluorescent fixture that uses 4 foot long T8 or T5 with a light diffuser shield that is sufficiently durable to withstand the routine cleaning/sanitizing. Select a fixture that allows for the installation of cold temperature tolerant electronic ballasts so that the fixture can be common throughout the plant for maintenance purposes.

- In order to illuminate the plant to the target levels, a 6 lamp, 4 lamp and 2 lamp fixture will be required. A fixture that can hold 6 lamps can also be fitted with 4 or 2 lamps but for some locations, a smaller specifically sized to hold 4 or 2 lamps might be the most suitable and appropriate.

- Develop a customized lighting layout that will provide the target lighting levels in the work areas throughout the plant. A computer generated uniform grid layout will not optimize lighting levels due to shadows from the overhead equipment so a customized layout will be necessary.

- The layout should also consider the need for task lighting in quality inspection and critical areas. A 2 lamp fixture is often suitable for task lighting installed just above the critical work level. Optimized task lighting is generally located closer to the critical work areas and will reduce the need for the lighting installed at the high bay level.

- Purchase fixtures with a 6 foot pig tail so it can be moved in the future as floor layout changes slightly or additional equipment is installed.

- Lamp lifetime is now upwards of 20,000 hours so it is important to select high efficiency electronic ballast that also has a sufficiently long projected life. Ballast service life is an important factor because some are plagued with early failure.

Hot Water

The KEMCO hot water heaters are designed and installed to operate at high efficiency but are not operating at optimum as indicated by the excessively high exhaust stack temperatures. The direct contact operation provides instant and continuous hot water and a design stack temperature of about $70^{\circ}F$ per the KEMCO website. The stack temperature should be no higher than about $20^{\circ}F$ above the incoming water temperature. The direct contact of the incoming water with the flame is certainly an efficient hot water heater design concept compared, for example, to a steam boiler fired hot water heater.

- These two heaters use approximately 5.5 million and 9.5 million BTU/hour based on their rated operation and the excessively high exhaust stack temperature indicates an excessive loss of 250,000 BTU/hr total for both compared to optimize system operation.

- The excessively high stack gas indicates that the heat transfer is not occurring as designed. This is due to an insufficient quantity and/or dispersion of the water entering the unit or a build-up of hard water residue on the stainless steel heat transfer media in the upper chamber.

- According to the manufacturer, removal of the build-up requires removal of the stainless steel heat transfer media from the heat chamber. It is possible that the lower layers of the media are heavily coated and removal from the chamber may cause breakage of some of the stainless steel pieces. One solution is to purchase a replacement batch of media so that the maintenance can be expedited. The coated media pieces can be cleaned with an acid solution that removes the layer of solids or some other appropriate cleaning technology.

- The estimated energy wasted up the stacks of the two KEMCO units is in the 2-3% range, which corresponds to savings in the $5,000 - $10,000 per year range.

Freezer Thermal Insulation

Data logging temperature sensors were placed in the -25°F freezer from August 18 through September 21. The sensors were fastened or placed to measure the freezer air space. No sensors were placed inside the walls or roofing material to obtain a temperature gradient in the walls or roof. We used 10 Lascar EL-USB-1 button battery powered temperature dataloggers to obtain the data.

Although the sensors indicated temperature excursions from minus 18oF to about 0oF, they did not indicate a diurnal variation that was influenced by solar loading on the building throughout the day. The data does not indicate that the temperature in the freezer is influenced by solar loads on the roof above the freezer. We did not examine the reasons for the temperature excursions but did notice that they appear to be repeated on a regular day to day basis. Perhaps they are due to routine production and shipping operations or may be related to the refrigeration system operation cycles. It could be that the high air circulation rates in the freezer keeps a uniform temperature throughout the air space in the freezer and solar roof loads are not apparent or measurable in the air space.

Thermal insulation is usually very cost effective especially considering today's energy costs. Installing additional thermal insulation to increase the roof and walls to a value of R50 would likely provide an attractive return on investment.

Infrared imaging may be a valuable tool to identify the need for additional insulation in the walls and roof. IR imaging may be helpful to identify a particular failure of the insulation or a air leak or thermal short circuiting but it may be more difficult to apply these results to broad thermal insulation and construction related issues. Another approach to identify the potential value of additional thermal insulation would be to install subsurface temperature sensors in the walls and roof in order to determine the temperature gradient through the structure and existing insulation.

ASSESSMENT W: POLLUTION PREVENTION/ENERGY ASSESSMENT FOR A LOCAL CITY (BY TECHSOLVE)

Related Chapters: Illumination, Documents, Solid Waste Management

A walkthrough of a City Hall, Service and Safety facilities was conducted. The city has 60 full-time employees. The following discussion provides current practices and recommendations for waste reduction/best practices.

The primary wastes generated from this facility are office-type wastes. Information on the type of wastes generated at this facility was provided. The wastes generated include:

1. Mixed office paper-recycled.

2. Used toner-recycled.

3. Used electronics-recycled/auction.

4. Cardboard-recycled.

5. Fluorescent lights—disposed.

Recommendations

One waste that could be reduced is office paper use by:

- Double-siding copies whenever possible;

- Using unneeded paper as scratch paper;

- Utilizing electronic correspondence whenever possible; and

- Electronic storage of drawings (environmental group reviewing option).

The City's facility can improve its energy efficiency through:

- Continued purchase of Energy Star appliances and other office equipment;

- Policy and posted reminders to turn off lights and computers when not in use;

- Placing copiers and printers in sleep mode when not in use;

- Occupancy sensors to reduce lighting costs; and

- Use of green tip fluorescent lights which contain less mercury.

Safety Facility

The Safety Facility houses the police and fire departments and the training room and Mayor's Court. Information on the type of wastes generated at the facility was obtained from facility personnel. The wastes generated at the facility include:

1. Office-type wastes-recycled.

2. Batteries—recycled.

3. Infectious waste-provided to Bethesda North Hospital.

4. Used electronic equipment—recycled.

5. Fluorescent lights-disposed.

Recommendations

The Safety facility could also reduce its paper consumption as described for City Hall. Because the facility operates 24 hours per day, the recommendations for turning off lights and computers do not apply. Purchase rechargeable batteries whenever possible to minimize battery waste.

Service Building

The Service Building houses the repair facilities and landscape/road treatment supplies. Information on the type of wastes generated at this facility was obtained. The facility generates the following wastes:

1. Used oil—recycled.

2.   Batteries—recycled.

3.   Antifreeze—recycled/reused.

4.   Freons—recycled.

5.   Salt storage—covered.

6.   Safety Kleen solvent bath—recycled.

7.   Used oil-recycled.

8.   Used oil filters-dispose.

9.   Fluorescent lights—disposed.

10. Tires-recycle.

11. Batteries-recycle.

12. Metals-recycled.

13. Grease pit residue—cleaned and disposed every year or so.

14. Absorbent—disposed.

15. Aerosol cans—emptied and disposed.

Recommendations

1.   The facility has a floor drain that runs down the length of the building that flows to a grease pit, which is cleaned of sludge every year. Oil and other spills are cleaned up using a —kitty litter ‖ type material. The facility needs to periodically reinforce the need for good housekeeping and spill prevention through use of catch pans so as to not have vehicle fluids discharged to the floor drain.

2.   Investigate the replacement of overhead T-12 fluorescent lights with the more energy efficient T-8 lights. The need for the overhead lights on bright, sunny days should be considered since the facility also has

natural lighting options. Apply task lighting through use of small lamps for tasks that require extra lighting and fine work.

3. Continue to ensure aerosol cans are emptied prior to disposal to avoid the generation of a hazardous waste.

City Energy Management Diagnostic

An Energy Management diagnostic was conducted to determine the current state of energy management practices for government buildings managed by the City. The diagnostic report attached provides 5 recommended action items for immediate attention:

1. Understanding opportunities through conduct of a technical audit to identify areas for energy improvement;

2. Energy supply evaluation for best cost;

3. Awareness and Training—general awareness training with employees and more extensive training for operations and maintenance staff;

4. Targets and key performance indicators for energy reduction; and

5. Energy trending to determine any problems that may arise.

Additional energy management activities are planned through the Ohio Department of Development's energy program. ODOD offers grant and loan funds to conduct a technical audit and development of a management plan. The anticipated savings for the city are 63,000 kwh per year and $6,800.

REFERENCE

[1]   University of Toledo web site (UT PPIS) website. Available from: http://www.eng.utoledo.edu/aprg/ppis/ppistools.htm (2012) [Accessed March 2012].

Index

www.ingramcontent.com/pod-product-compliance
Lightning Source LLC
Chambersburg PA
CBHW050838220326
41598CB00006B/391